A teacher's guide to

ANCIENT TECHNOLOGY

Jake Keen

ENGLISH HERITAGE

CONTENTS

Ancient technology is fun!

ABOUT THIS BOOK

The kettle boils and all round Britain people pour out cups of tea creating a massive but temporary call for electricity. On the kitchen shelves are foodstuffs and beverages, spices and condiments, imported from the far corners of the earth. Pick up ten objects at random from those around you. The chances are high that you will be holding things grown, manufactured or wrapped by unknown persons living lives remote from your own. Yet however similar or dissimilar to ours those lives may be, they are rooted in a past we all share. We wear clothes spun from fibres derived from oil by incomprehensible means. We operate the switches and buttons and keys of electronic devices whose workings are unfathomable to us and yet on which we depend. We drink from containers wrought of exquisitely fine metal, extracted from crushed rock and extruded at huge temperatures by the extravagant use of the same power' that boils the kettle, ... and then we throw them away. All the while, so frequently we no longer notice them, carriages with wings hurtle great numbers of our species across the skies at speeds literally unimaginable at the beginning of this century - above the lands we roamed on foot for more than 99% of human history.

A pupil in school today has never known any other world. Most have grown up observing that adults, although great users of the technology we all now share, are not able to build, repair, nor even understand large parts of it. Even the specialists, who service cars or take the backs off televisions and washing machines, usually do little more than replace failed components made far away by people who themselves could not assemble all the separate parts. To make a microwave oven, a drink can, or even a ballpoint pen from scratch – that would be a thing!

In a sense we grow up more helpless than ever before. Although collectively we have extraordinary – unprecedented – powers of control over the earth's resources, and have marvels of engineering available in every household, individually we have lost most of the skills common to our ancestors: building a home with family, friends or neighbours from local materials, the long but shared times producing and then preparing the raw stuff of food or clothing. This is not to say that life in the past was uncomplicated or idyllic. Technology, in our era, has followed the unswerving pursuit of easier and less time-consuming

unlikely a damaged blade would leave a household so impotent as one with a jammed video machine today.

For decades now, the majority of people in Britain, both adults and children, have spent more time watching television than doing any other single thing besides work or school and sleep. Television seduces us with the illusion that we are participating in the world, and we probably have more surrogate experiences than the richest medieval kings and queens had real ones. But though more wonders pass in front of our eyes than ever before, we actually make and do less and less first-hand.

First hand experience: Building in chalk cob. Cob walls are made by trampling a sticky mixture of clayey soil or chalk with a binding material (for example horse hair, straw, or hay) and building up in layers. Daub is the same mixture applied to a woven wood (wattle) framework as in wattle and daub.

ways of getting the required job done. However, in the process we have abandoned much knowledge and many skills. A child growing up in, say, the Iron Age, would know how its home was built, where to find the right woods, where to cut thatching materials, how string was made, how wool and flax were spun and then woven on the loom, where and how crops were best grown and how to tend and then slaughter the animals. The tool makers may have been itinerant bronze and iron smiths and their art may have appeared full of mystery or magic, but it is

This book is about teaching children in the TV age about the past by offering them the certainty of first-hand experience, enabling them to explore technology they can build, repair, fully understand and use. In common with other teacher's guides in this series it advocates practical, exciting and memorable projects, which can be linked directly to the National Curriculum, are truly cross-curricular and multi-disciplinary, and which are mainly designed to take place outside the classroom.

WHAT IS ANCIENT TECHNOLOGY?

Ancient technology, in the context of this book, means simply the skills, techniques and associated artifacts of the past, and refers to what is very old from a child's point of view: so both the clay pipe clamped in the jaws of great granddad in the old photo and the flint axe ploughed up in the field next to the school are examples of ancient technology. In practice, most of the examples will be drawn from prehistory, and the simpler devices and inventions of later periods.

Most people, children included, seem to be fascinated by traditional crafts especially if they can watch old fangled gadgets from the past being used. Handling objects from

English Heritage Education Service

Incorrect illustrations such as this one used to be in many school textbooks. It is based on a misunderstanding of archaeological evidence that interpreted grain storage pits as human dwellings. It took many years for the revised archaeological interpretation to reach school textbooks.

a previous time, or experimenting with past methods is a powerful way of introducing history, and can stimulate a strong academic interest in this and other subjects.

In part, learning about – and attempting to do – ancient technology is concerned with giving children back their roots: enabling them to become acquainted with the

natural materials from which all technology originally derived. However, it is also about learning how we see the world and about standing back to gain a perspective on our own times.

OBJECTIVE INTERPRETATION

Any attempt to re-create or re-construct the past in a completely objective way is an impossible ambition. However, making the attempt can teach us about prejudices held in the present. Such value judgments, often held unconsciously, unwittingly colour the way we interpret evidence from the past. Here are some examples.

Manual work is inferior to intellectual work

This long established view seems to be built into our educational system. For example, in 1605 Francis Bacon noted 'it is esteemed a kinde of dishonour unto learning to descend to Enquirie or Meditation upon matters mechanicall'. And yet it is the actual experience of tackling practical problems which, more than anything else, leads to innovation in technology. Both head and hands need educating.

Human societies evolve, just as plants and animals have evolved

Darwin's theory of evolution profoundly changed the way people subsequently understood the world. Until his time, it was commonly believed that the creation of the world had occurred in 4004 BC. As it became clear that not only the earth, but people too, had a prehistory, scholars began to study human development. Soon models of the emergence of human society emerged, placing Europeans, who at this time ruled much of the world, at the pinnacle of development and other people at various stages along

the road to progress.

The view of early anthropologists that humans evolved from a state of savagery (primitive hunting) through barbarism (simple farming) to civilisation (advanced technology and writing), was based on a link between cultural and biological evolution. Any such link is now largely discredited, but its legacy remains. The theory influenced Marx and Engels, and has been used to legitimize the view that some cultures and the people associated with them are less developed/significant/valuable than others. For example, the extreme anti-semitic policies of Nazi Germany were built on the notion that some people are more human than others. But even in small ways our picture of the past was dominated for years by this view. The illustration on this page from a history book of 1958 shows a pit dwelling supposedly occupied by Iron Age people in Britain. When archaeologists discovered large numbers of chalk pits containing broken pieces of pottery and chewed bones, they assumed our ancestors, conforming to the above theory of human development, lived in squalor in holes in the ground. It took decades for the realisation that these were in fact grain storage pits, not dwellings, to reach the authors of school text books.

People using simple technology were (and are) primitive and backward

No evidence has been discovered to support the view that for at least the last 40,000 years our ancestors were any less intelligent or capable than ourselves. They lived in different times and had access to different technology. We live with the accumulated discoveries and inventions of many thousands of years. Where isolated groups of people using simple technology, such as the stone tool cultures in some of the remote valleys of New Guinea, have come into contact with modern technology this century, it has only taken them a generation to produce airline pilots and computer programmers of their own.

Technological progress makes modern people advanced and civilized

Possessing sophisticated technological labour-saving devices does not in itself convey upon the owners any understanding or competence in the technology concerned. While it is undoubtedly true that one technological breakthrough can lead to another, no overall pattern of human evolution can be attributed solely to technological development.

General Pitt-Rivers, one of the 'fathers' of scientific archaeology, was another nineteenth century researcher influenced by Darwin. He set out to chart the evolution of technology and travelled the world collecting artifacts. His remarkable collection in Oxford, demonstrates the ingenuity and inventiveness of indigenous people in making use of the natural environment to satisfy their technological needs. However, it does not reveal the overall pattern of evolutionary development for which he was looking.

Modern technology is dangerous and can threaten livelihood and health

In the early nineteenth century the Luddites were convinced that new machines being introduced into factories would destroy jobs and cause great hardship. Machinery was smashed, the perpetrators were transported or hanged but the movement did little to slow down what was regarded as the race towards progress. Mass production with reduced labour costs made for cheaper goods, increased sales, and substantial profits for investors. Then, as now, the social and environmental costs of such progress were not always considered.

Although today we are more aware of the local and global effects of industrialization, and more fearful of some of the unforseen consequences of pollution, we remain both dazzled by new technology and forgetful of life before it. The pain and fear of illnesses, now eradicated, and the servitude to domestic chores, now consigned to machinery, have been quickly forgotten.

WHY TEACH ANCIENT TECHNOLOGY?

It is relevant to how children learn

Children learn best by doing: *Tell me something and I'll forget it; Show me something and I'll remember it; Let me do something and I'll understand it.* The truth of this old proverb has been borne out by recent research:

% Information retained after:	3 hours	3 days	3 weeks
Reading	50	5	3
Hearing	70	10	5
Seeing	75	25	10
Hearing and seeing	85	65	50
Doing (active participation)	95	85	70

Various sources

Studies of the way infants acquire language have revealed that the process doesn't work properly unless the child has something meaningful it wants to talk about – such as its adventures exploring the tangible world. Many children seem to have an affinity for natural materials: sparkling stones, wet clay, springy wood, stray feathers. Given the chance most delight in making bows and arrows, constructing and decorating shelters, conjuring up camp fires and cooking over them.

Cranborne First School

It is relevant to what children are supposed to learn

While neither ancient technology nor archaeology, on which much of our knowledge of it is based, are subjects in the National

Curriculum, both can provide a relevant context for practical work in a wide variety of curriculum areas and at different key stages. Cross-curricular work is often recommended but it is hard to organise in practice in many schools. Ancient technology projects, whether they start from initiatives in science, history, geography, art or technology, stimulate interest in a variety of disciplines (see pages 32 to 34).

The need for physical education

The beginning of state education was in part a response to the perceived need to protect children from excessive exploitation and in part an attempt to raise standards of health and morality. Now, five generations later, most young children, when suitably challenged, relish the opportunity for hard, physical work. Today, probably for the first time in human history, such activity is not part of the ordinary experience of growing up. Recent research into incipient heart disease implicates lack of exercise in young people, despite their compulsory quota of PE and games in the school timetable. By reacting, more than a century ago, to the iniquity of forced child labour children are

It took many years before schools could record full attendances because the help children could give in certain types of work was still seen as necessary - as in this rural example.

now obliged to spend ten years, mostly sitting down, in schools. The National Curriculum can accommodate learning through practical tasks which simultaneously challenge mind and body.

SOURCES OF INFORMATION

There are a number of books that contain information about aspects of ancient technology and there are even video films which show attempts to recreate old techniques (see *Bibliography and resources*). A satisfactory project could be conducted using such secondary sources although first-hand investigation offers the excitement of studying primary evidence. Ancient technology can be discovered through material remains, records and other sources.

MATERIAL REMAINS

Archaeology has been defined as the study of people through their material remains. It is essential to know the context – where and in relation to what else – in which remains are found as these relationships can give vital extra information. These remains can range from grand structures,

English Heritage Photographic Library

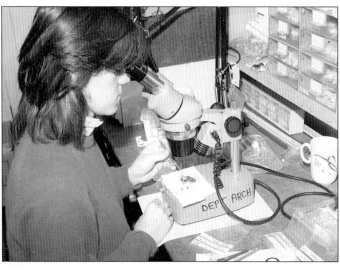

English Heritage Education Service

Two levels of archaeological remains: an aerial photograph of the massive hill fort of Maiden Castle in Dorset and an archaeological scientist studying microscopic food remains from excavation.

Many museums now illustrate the context of archaeological finds by creating a full-scale model of a vertical section through a site with artefacts and ecofacts sticking out of the side at appropriate levels. Constructing such a model may be a project that could be done in conjunction with a local museum or archaeological unit many of which have lending collections of less important finds not destined for display or further research.

apparently built for religious or defensive purposes, such as the prehistoric stone circles at Avebury or the huge Iron Age hillfort of Maiden Castle, to the mundane refuse from day-to-day human activities. In fact, such rubbish – food remains, broken pottery, discarded tools and worn out utensils – constitutes most of

the material recovered by archaeologists. From these artifacts (objects used, modified or made by people) archaeologists make inferences about the original owners and their lifestyles. How much information archaeologists can get from such finds depends on a range of factors including when it was thrown away, the kinds

of material from which it was made, the type of soil that has formed around it, and how much the ground has been disturbed subsequently. In general, the greater the time that has elapsed the less well preserved the material evidence will be, although certain materials such as stone, pottery, metals and bone tend to last better an wood, leather and textiles. here are conditions, however, in aterlogged soil or under water, here organic material can rvive more or less intact, and ese cases, together with the rare stances of preservation due to treme cold or dryness, can ovide a disproportionate antity and quality of formation. Archaeologists try use any information that might lp them understand what life was e in the past including ecofacts on-artefactual, organic and vironmental finds such as sect remains) that can help eate a picture of the climate d environmental conditions the past.

Apart from exhibiting the choice ds from excavations, museums so often contain artefacts taken om ancient sites before the days scientific archaeology, when a vourite pastime of country ntlemen was digging into burial ounds to retrieve grave goods. ch digging destroyed the context the finds and a great deal of formation was lost for ever, though collections gathered in is way were the foundation of any of today's museums. It ould be stressed to pupils that e value of archaeology is in ncovering clues that help us derstand the past. Most think value in monetary terms and agine that archaeologists are ying to locate and dig up tiquities because they are orth lots of money.

Other artefacts in museums ve never been in the ground and ve been looked after carefully, metimes for centuries. They may e of fragile materials which would ot have survived in the soil. Such ecial items, together with llections of bygones and old ols can give valuable insight into mestic and country crafts, some which have survived virtually nchanged from prehistoric times.

RECORDS

useful definition of prehistory is e study of the past before records writing began. From then on rious kinds of recorded evidence

An illustration from the sixteenth century book *De Re Metallica* by Agricola showing the rag and chain system used for draining mines. Contemporary illustrations such as this are very useful in understanding early technology.

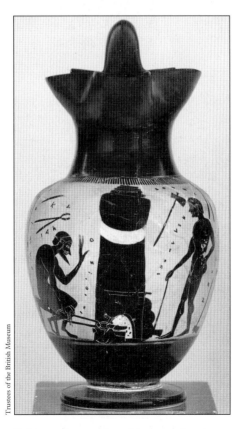

Evidence for metal working on a Greek vase made in Athens c.510-500 BC. The man on the left seems to be removing a piece of heated metal from the furnace prior to shaping it on an anvil. The younger man (possibly an apprentice) waits with a hammer.

can be added to that produced by archaeology and the picture of the past becomes increasingly clear towards the present. Concerning ancient technology, certain kinds of record are particularly useful.

Written
There are passing references to technology in some of the earliest surviving texts, giving us clues to the practicalities of life long ago. In 401 BC Xenophon, leading his Greek mercenaries through the wilds of Kurdistan in winter, came to a village. 'The houses here were built underground; the entrances were like wells, but they broadened out lower down. There were tunnels dug in the ground for the animals, while the men went down by ladder'.

Technical manuals, such as that produced in Germany by Agricola in 1556, can be very useful in showing the workings of the tools and machines of later centuries, through a combination of words and drawings.

Pictorial
Depictions of aspects of daily life can be amongst the most valuable clues to understanding ancient technology. They include engravings on rock, bone or antler, wall paintings, decorated vases and illuminated manuscripts. Background detail in the paintings (or study drawings) from the fourteenth century onwards enable us to see through the eyes of more or less careful observers from the past.

Three Dimensional
Sometimes contemporary models or carvings can help us understand the way artefacts were used, something not always apparent in an archaeological context.

Photographic
Early photographs and films captured people using methods now extinct, some of which had been in existence for centuries. The first ethnographic films, too, have influenced the way material remains have been interpreted.

Wooden funery model of brickmakers. Middle Kingdom c. 2,000 BC. A squatting figure moulds clay into bricks in a wooden mould. His companions mix clay for more bricks. Length 32.7cm.

OTHER SOURCES

Ethnographic parallels

Ethnography, the study of the material, social and linguistic characteristics of contemporary ethnic groups, can help us understand the technology of the past. For example, circles of postholes (ie where posts once stood) uncovered by archaeologists are recognised as being the remains of houses partly by analogy with modern round houses lived in by people in Africa and other parts of the world. Wedge-shaped stone objects from early farming sites in Europe were identified as axe-heads partly because similar tools were being used for tree felling in South America when the objects were first excavated. This interpretation has since been reinforced by the discovery of prehistoric axes with wooden handles intact, and by experimental work which has demonstrated the effectiveness of such stone tools in felling. The main danger in using ethnographic sources comes from assuming that people living in non-industrialised societies, and using artefacts similar to those used by our ancestors, must hold similar beliefs and values. The assumption *may* be correct but there is almost always insufficient evidence to support or deny it.

Experiments in growing and storing ancient crops: harvesting wheat with sickles; filling grain storage pit; sealing pit with wet clay.

Domestic and country crafts

Many existing crafts, such as basketry, hurdle making, spinning, weaving and flint knapping have their roots in prehistory. Styles and techniques have undoubtedly been modified over the centuries, but basic techniques have been the same for very long periods. The revival of interest in such traditions has been manifested by the growth of craft shows and exhibitions. Expert practitioners in all parts of the country are often willing to share their skills or advice to help a school group get started.

Re-enactment groups

The flourishing groups which specialise in re-enacting the battles or daily business of different historical periods are another expression of contemporary recreational interest in our past. Individual members are enthusiasts, often skilled and knowledgable about the technology of their period. Some groups regularly work in schools.

Experimental archaeology

Many of the activities which will be suggested in the following pages could be referred to, more or less strictly, as experimental archaeology. The class illustrated on page 33 were 'experimenting' to find out how to make and paddle coracles. However, the testing of a hypothesis by repeated trials, controlling one variable at a time, can not only introduce the idea of a 'fair test' in an exciting way, but can also be of genuine interest to archaeologists trying to find out how ancient technology might have worked. Specialist groups researching ancient technology are keen to encourage such work.

PRACTICAL PROJECTS: WHERE TO START

Some of the activities suggested in this book are suitable for Key Stage 1 pupils and can be safely conducted by responsible adults with no previous experience and relatively little preparation. Others are much more ambitious and would only be suitable for a school or group wishing to commit itself to a great deal of industry and research. The rewards tend to match the time and enthusiasm invested and although many of the projects are unorthodox, and sometimes require scouting for elusive advice and strange materials, they can generate a great deal of support and goodwill from parents and the local community. It is refreshing for pupils, parents and teachers to be able to share learning experiences on a level footing and with equal interest.

An ancient technology project introduces pupils to a wide range of skills. Here pupils are cutting reeds to be used for thatching.

USE OF SCHOOL GROUNDS

There is sufficient space in most school grounds to develop ancient technology projects. It may not be possible in all circumstances to attempt some of the larger activities but something can be done even in the most cramped of urban environments. A legitimate worry, especially in urban areas is vandalism. Chickerell Primary in Dorset tackled this head on by involving the children in the issues (see page 12). The result has been a strengthening of community involvement in the school. Even if there is a risk of damage by outsiders, two questions need to be asked:

Is the work done on the project sufficient outcome in itself, or is the finished structure the primary objective?

■ How should children be taught to respond to vandalism if and when it occurs?

Where it is considered prudent not to embark on ambitious or vulnerable projects resulting in permanent structures, there are many alternatives – experiments that can be completed and cleared away in a session or portable work that can be stored in the classroom.

ORGANISATION AND MANAGEMENT OF ACTIVITIES

Offering pupils the chance to participate in stimulating and challenging activities outside the classroom can lead to high levels of excitement. This should not be a reason for avoiding such work, but prior planning is essential. There are a few basic points to bear in mind:

■ If you are working with a whole class there are very few projects that require every pupil to do an identical task. It will be normal, therefore, to work with a number of smaller groups (ideally no larger than eight). These must be organised so that every pupil has a meaningful job. A recipe for disaster is to permit a group of frustrated pupils to watch the lucky ones doing the interesting work.

■ The smaller groups will need supervision. Parents or grandparents can normally be recruited quite easily. Brief them fully and carefully as to their role beforehand, get them to practice if necessary, and arrange for them to stay with an activity even if pupils change theirs.

■ Most of the activities suggested in this book could not be completed in a single lesson.

All necessary safety precautions must always be taken.

A special afternoon or day would be required for most of them. Longer projects, involving ongoing work could be organised differently, with small supervised groups being released in turn while the majority of pupils continue normal classroom work. This can be a great motivator and assist in achieving the standards expected from the pupils. Much of the Chickerell work was done in this way.

■ Trialling of the activities beforehand is essential to ensure risks are properly assessed and appropriate tools and resources provided.

■ Some projects will generate so much interest that pupils may actually want to stay after school. This can be very rewarding and help develop an excellent relationship with a class, but can also, understandably, be impossible for many teachers. An after-school club, however, may draw support from your local museum or archaeological unit. Particularly keen pupils may be able to participate in some 'real' archaeology by helping to wash pottery sherds from local excavations, or assist in field walking. Some pupils will soon become proficient at weaving

and other crafts and, like the expert net makers from Bembridge (see page 14), become a great asset to their school.

SAFETY

All the activities mentioned in this book derive from times when personal safety was a matter of individual, not corporate, concern. Protective goggles were certainly not worn by flint knappers in prehistory, nor was the safety of children working in the textile mills in the nineteenth century much considered. In 1815 one per cent of cotton operatives died yearly as a result of injury or disease.

It is clearly of paramount importance that no pupil should be put in any danger. For this reason, before starting any ancient technology project it is essential to assess the risks and decide how to reduce them to an acceptable minimum. For example, if a process is potentially very dusty or smoky it will be necessary to establish in advance how pupils will be able to avoid these hazards. Because guidelines for traditional ways of working do not exist, all techniques should be tried and tested by teachers before introducing them to pupils. It is well known that most younger pupils will jostle to be at the front to watch something that fascinates them. Until quite late in their school career, pupils are not good at noticing the consequences of their actions on others. Carrying long poles or using sharp tools, for example, must be organised with this in mind. It is the process of such organisation that helps to teach sensitivity and awareness of others.

Understanding the nature of materials and how to use them is both the subject of this book and an area of diminishing knowledge: which woods spring back and which can suddenly splinter, which juices stain and which are toxic, which stone shatters on the fire, which remains whole. Although what was once common knowledge is now held by only a few, these individuals, sometimes cast to their surprise

in the role of 'expert', have often published or are accessible through their efforts to keep the old traditions alive. Most value and are willing to share what has been passed on to them (see *Bibliography and resources*).

INTRODUCING ANCIENT TECHNOLOGY

Starting from evidence

A look at what is lying on the surface of the soil, a visit to a local archaeological site or museum, or a detailed study of maps or reports of excavations in the area could all lead in to a project to reconstruct a prehistoric building, a kiln, or the process by which a given artefact (a flint tool, a linen bonnet, a plaited loaf) was made at a particular period in the past.

Problem solving

Give your pupils some basic information about the materials known to have been available at a particular time in the past and then give them the results of an archaeological investigation: for example, a particular pattern of postholes was discovered at the Iron Age hillfort of Danebury. Get your pupils to make a model of a structure that would withstand strong winds and keep out the rain using posts arranged in the ground in the same way. A more open ended task would be for them to imagine they are hunter-gatherers stopping for a while near rich hunting ground. They should make a shelter that will keep out the rain for a few days from wood cut nearby (provided). The materials chosen to cover the roof (hides, grass, leaves, turf, bark?) will affect the design of the shelter. Building the structure can be relatively simple. Roofing it can be more costly in time and materials and access to materials may constrain a school project just as it did hunters in Mesolithic times. (A decision should be made in advance whether modern materials can be allowed for keeping the rain out.)

Evidence from the earth: pupils enjoy excavating a small area of the school grounds taking care to record, clean and exhibit their finds.

Jake Keen

Jake Keen

STRATEGIES FOR STARTING

Evidence from the earth

The aim of this exercise is to show younger pupils (Years 2, 3, and 4) that the soil is full of interest and contains clues that tell us about the past. The purpose is not to encourage archaeological prospecting so a patch of ground already selected for planting or making a garden or pond would be best. If, by chance, pupils discover

Another approach may ignore archaeology altogether and draw on ethnographic evidence. For example, people from many cultures have used woven grasses or bark to make containers. Get your pupils to design and make a bag, box or basket to carry fire, eggs, grain or water from the materials provided. The work generated may be very successful as technology, but care should be taken not to identify the results as belonging to a particular period unless there is specific evidence to link the techniques involved with that time.

Experimental and investigative science

Hypotheses about how pots were tempered and fired, ores melted, bread baked or glues distilled in the past can be tested experimentally. Such

experimental archaeology has the advantage of being fresh (frequently answers are not known); concerned with essential issues such as food and shelter; and are of potential value in developing a better understanding of the past.

It is also an effective means of teaching essential aspects of science.

Empathy

Drama, role play, creative writing, and storytelling as well as art and music can use history or prehistory to stimulate creative work. Introducing a practical element of ancient technology can provide a greater sense of realism to help focus the imagination.

> ## IRON AGE POEM
>
> *It's uncomfy on this wooden bench,*
> *But it's all we've got,*
> *Our ancient house is bent and cold,*
> *And the walls begin to rot,*
> *But still we love our little house,*
> *It overlooks the lake,*
> *So now we can go fishing for*
> *non-starvation's sake,*
> *We don't always get food from water,*
> *Sometimes a deer we like to slaughter,*
> *We roast it on an open fire,*
> *The smoke grew higher and higher.*
> *This is the end of our description*
> *And let me assure you it's not fiction.*
>
> by Stephen Hill and Joe Wilsdon

anything immediately identifiable as real archaeology, they should stop work immediately and call in

Chickerell County Primary School: Iron Age Roundhouse Museum

The Iron Age roundhouse project, stimulated by a visit to Cranborne Ancient Technology Centre, was one aspect of a broader, cross-curricula, term-long Year 6 theme on 'Structures'. During the initial day-long project pupils were fully engaged in decision making and problem solving, turning ideas into reality.

The project addressed three fundamental issues:

■ the children's social and personal skills and attitudes

■ an understanding of history through an investigative and problem solving approach

■ a development of community and industrial understanding.

The day was divided into a series of workshops. The aim was to build a three metre diameter roundhouse, which would be functional, providing a learning resource for the school and community.

■ coppicing at Garston Wood, RSPB Nature Reserve

■ building the main structure of the roundhouse with the coppiced wood

■ cutting reeds at Radipole Lake to enhance the environment for the lake's wildlife and to provide vital thatching material and

■ thatching the roof – design and methods of making the house waterproof.

The pupils were keen to complete the plastering of the walls with clay and straw during their holidays immediately following the project.

As part of the project follow-up the Dorset Theatre in Education Team spent a day with the pupils in role play discovering what it may have been like for the Durotriges during this period in history.

In the next year the incoming Year 6 pupils took over responsibility for the house. They were especially concerned with a stream of unwelcome visitors and vandalism to the house.

A local industry/school liaison officer, worked with Year 6 to support their campaign against vandalism, and to develop an appreciation of the world of commerce. Small groups of pupils took responsibility to interview and meet members of the community, school governors, parish councillors, teenagers and local youth leaders to publicise their work in the community and to gain respect.

It was decided to have a public launch day for the roundhouse that would include a museum display on 'A day in the life of a Celt'. A stockade was erected around the house, complete with a kiln, oven, weaving looms and storage pit.

Through the practical and oracy skills developed by the industry/school liaison officer the pupils were made aware of the factors involved in running a business. Links were forged between pupils and Sergeant Bun's Bakery who showed them how a small family business is run. The bakery also supported pupils by providing bread for the museum and cakes and rolls for the launch day.

The pupils, prepared for the launch day by making decisions about the programme of events, guest lists and organisation of the day. The day was to be a celebration of events past and present, including demonstrations of ancient craft and lifestyle.

The pupils invited those who had helped with the project including parish councillors, youth leaders, previous Year 6 pupils who had started the programme, school governors, the Mayor and Mayoress of Weymouth, parents and friends. The event concluded with a 'Handing Over Ceremony' by the children of the responsibility for the museum and roundhouse to the incoming Year 6 class.

The commitment and enjoyment shared by the classes brought a great sense of achievement and through the practical approach made history become alive and meaningful. The pupils have formed many partnerships outside the confines of school. They have gained many skills, self confidence, self esteem, self maturity as well as a knowledge and understanding of many subject areas.

professional archaeologists to look at the site.

Mark out a given area (say 500 x 500mm for each group of four pupils) lift the turf, and ask the pupils to scrape away the soil in horizontal layers saving everything that may have been made, altered or introduced by people and anything else that looks interesting. A margarine tub is ideal for containing the finds which should be cleaned (using old toothbrushes in a tray of water) once subsoil or a sensible depth has been reached. Sieving the soil (5mm mesh) and hosing earth off the finds is a big help in damp or clayey conditions.

Most pupils have never looked closely at or in the ground and find this activity fascinating, but often have difficulty in identifying what they have found. That does not matter over much. The cleaned objects should be sorted – by whatever criteria the pupils and teacher together think appropriate – by material/colour, age/origin, natural/human made... Few school will be built on completely sterile land and even in the most unpromising spot it is extraordinary how much debris

Dear Contact

In my capacity of Youth Leader, I was delighted to accept an invitation from Class Y6 to visit them, at Chickerell School, to find out more about their famous Round House. I was met at the gate by Michael and Ian, then taken to meet two more Michaels, these four then took me on a very interesting tour of their classroom, where I looked at photos and read pieces of work written by the children about their Round House project. I then was taken to see it, by my young guides, and was shown the damage that had been done to the building, which greatly saddened me.

From there I was escorted to the staff room where my young hosts plied me with coffee and biscuits. We then discussed ways of preventing any further vandalism. I agreed to ask the members of the Youth Club for their suggestions. This resulted in some of the older members offering to "keep an eye on the hut" and three lads - Scott Mowlam, Justin Palmer and Danny Allen - volunteering to help with some of the repair work.

[...] "Contact" and as [...] rents to speak with their [...] this letter. In [...] have what [...] the member [...] issues, beh [...] at home.

[...] increase in [...] to take p [...] ed by t [...] an exc [...] ing to [...] ho [...] "Co

Pupils from Chickerell School tried to counter vandalism by working with the local council and other groups (see opposite).

Chickerell Parish Council
Ms Elizabeth Symonds –
Clerk to the Council

Pupils of Year 6
Chickerell Primary School
Chickerell
Weymouth
Dorset.

Dear Pupils of Year 6

Thank you for your letter dated 9 March 1993 letting us know that vandals have damaged your replica of an Iron Age Round House.

The Parish Council were extremely sorry to think that there are people in Chickerell who are so thoughtless and uncaring to have caused this damage. We are appealing, through Contact Magazine, to all residents in Chickerell to help us find out who these people are, then hopefully we can put a stop to this damage.

We would like to ask you if there are any other ways in which you think the Parish Council could help you stop this vandalism. If there are, will you please write and let us know.

Yours sincerely

Clerk to the Parish Council

Dear Householder

The pupils of Class Six would like to tell you about our Iron Age Roundhouse, which was built by last year's Class Six. We plan to continue the project and have been busy building the surrounding fence.

We intend to make our Roundhouse into a mini museum, which will be used by other Schools and the public.

Unfortunately our Roundhouse is being destroyed by vandals. We have got a major problem and intend to stop it. Our evidence is litter which includes beer cans, cigarette ends, chewing gum etc. They have also climbed up the side of our Roundhouse. The ring which has been holding the reeds in place has been broken in two.

If you have any ideas to stop this vandalism, please contact Class Six, Chickerell CP School.

Thank you

Class Six

[...]eft by human activity will be [un]covered. Pupils can work out [te]sts to discover what unidentified [o]bjects are made from or ask adults [t]o help in their identification. Using a tea strainer to wash the [in]ner contents of a handful of soil [c]an reveal tiny snails and seeds.

By studying such finds environmental archaeologists can discover changing patterns of landscape use over the centuries.

This exercise can teach about local history, geology, natural history, how soil is formed and can stimulate creative writing if the objects discovered are linked together in an imaginative way.

A load of rubbish

Modern refuse can be used (with suitable precautions for safety and hygiene) to look at the very recent past and to discover the sort of

Forelands Middle School, Bembridge: living history

Living history events can provide a valuable learning experience for children, but they can also stimulate and challenge teachers as well. In 1989 I took a group of children to participate in a medieval fair at Carisbrooke Castle as a group of medieval warreners. We tanned rabbit skins, made rabbit nets, and some of the boys demonstrated their ferrets. After the event I realised I knew nothing of the processes involved in making twine or rope, nor the fibres that would have been used.

A great deal of research into the subject followed, with hand operated equipment being made to carry out the processes discovered. These processes were introduced to children during history and technology lessons, and now children from this school regularly demonstrate at events put on by re-enactment societies, visit other schools to demonstrate specific skills, as well as make a range of useful products both for themselves and for sale.

Rural rope works were found

Pupils soon become experts: here one pupil concentrates on making a tennis net from natural fibres.

throughout the country, serving the needs of farmers for sheep nets, wagon ropes, tarpaulins, plough lines, rabbit nets and so on. On the coast the requirement was for fishing nets, ropes for ships` rigging and pot lines.

Activities at school attempt to emulate this industry. Flax is grown and retted, broken,

scutched, hackled, and finally spun into yarn which can then be made into netting twine. Netted products include teddy hammocks and this summer a patiently made tennis net! Ropes are made from natural synthetic yarns with uses as diverse as school skipping ropes, dinghy painters, and lobster pot lines.

An activity such as this can be involved in many aspects of the school curriculum. After a visit to "Mary Rose" in Portsmouth we made a replica nine-stranded anchor cable, now in use as the school tug-of-war rope. Studies of the Industrial Revolution are made more real to the children with experience of hand textile processes. World trade and the part played by developing countries becomes relevant in terms of coir, hemp, and jute production and the effects the use of synthetic fibres has on these countries; and almost any aspect of the work of "Forelands Twine, Net and Rope Manufactory" involves the use of science, maths and technology.

Nigel Tibbutt

information that the older rubbish excavated by archaeologists can divulge. At the beginning of a school year each member of the class could be asked to bring in ten typical (washed, cleaned or bagged) items from the dustbin at home. On a suitable surface the 300 or so separate pieces of rubbish could then be analysed according to what they are made of, country of origin, and proportion of each type (for example, packaging/food waste/ the result of cleaning/breakages/ reading material/other). What can the pupils say about the group whose rubbish it is? If there is time the rubbish could all be buried in a suitable pit and covered over. At the end of the year or term the contents could be carefully disinterred to find out what still remained and the effects of rotting on different substances.

The contents of a modern bin can provide lots of information about the people who threw away the rubbish!

STRUCTURES

Shelter is one of the essentials for sustaining human life and young children frequently make camps and dens when they are playing. An important part of this play involves getting inside the shelter, so although making models of structures can be enjoyable and instructive, they do not have the same appeal as 'real' shelters. The great advantage of constructing buildings in an ancient style in a controlled learning environment

materials and the stresses, loads, and forces acting on structures, builders tended to be quite conservative in their designs, sticking to forms and techniques

of medieval churches and cathedrals collapsed due to overambitious design.

Apart from tents and caves and holes burrowed in the ground

A building with load bearing walls.

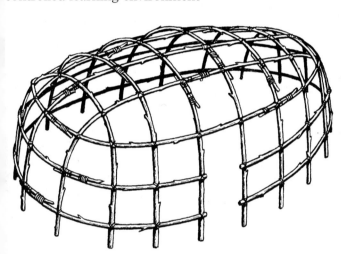

A tensile framed building.

is that it is safe, matches children's interests and provides an ideal opportunity for teaching about materials, structures and forces and building traditions.

In this book a structure is considered to be any assemblage of materials which is intended to sustain loads. Structures can fail for a number of reasons:

■ if poorly designed (incorrect shape/inappropriate materials)

■ if the joints connecting component parts fail

■ fatigue – caused by loads varying over a period/vibrations

■ excessive loads.

Before scientific means of calculating the strength of

which were tried and tested. Superstition probably played a part in upholding traditions, and ceremonies such as the laying of foundation stones and the christening of ships with champagne are the final vestiges of some very old rites. Sacrifices and their associated ceremonies, offered to placate whatever powers might cause a structure to collapse, were presumably once the only kind of insurance policy. This is not to say that traditions did not change nor that spectacular failures did not happen – for example, a number

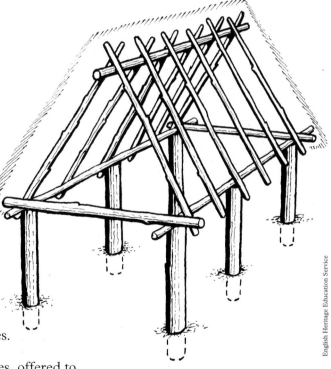

A post and beam building

to make living quarters ancient or traditional buildings seem to have been constructed in three main ways.

15

■ Tensile or bent frame with covering – ie where there is no distinction between walls and roof. The dome-shaped frame gains enough springy strength from its bent saplings to support a covering of hide or thatch.

■ Post and beam (joined) wood frame with various walling materials. The roofs are usually of straight poles lashed or otherwise joined to the frame.

■ Structures with load bearing walls (compression shell). The building material serves both to form the walls and to support the roof. These structures generally leave little or no trace in the ground.

These three types of structure, which can be domed, conical or rectilinear, may be combined in various ways. They can be built of a wide variety of natural materials. Clay, mud, turf, stone, timber, withes (6mm rods of hazel, birch or willow), bark, reed, grass, hide, and even mammoth and whale bones have been used, as well as compacted snow. The materials used reflect what is available in the local environment, but also what weather and climatic conditions the building needs to withstand.

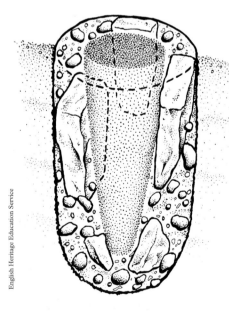

English Heritage Education Service

The cross section of a post hole. This section shows that the original post was round, pointed and kept secure by three packing stones.

Gathering materials and woodland management

A project such as building a roundhouse is a practical way of showing pupils how extraordinary, rich, varied and valuable nature is while, at the same time, giving them a personal stake in its conservation.

In effect, nature in modern Britain means habitats of one kind or another that have been managed in various ways. There is hardly any land unaffected by human action left. However, there are considerable areas, many of them now protected in some way, where the habitat is still looked after – or managed – to maintain the balance of wildlife that currently exists. The ideal way to do this is in the traditional manner, which is where ancient technology comes in.

Most of the old woodland and countryside crafts, many of them probably surviving for thousands of years, have suffered a drastic decline this century as modern technology has replaced their end products with cheaper, longer lasting, or simply more fashionable alternatives. For example, the regular coppicing of woodland and cutting of reed for thatching, have been abandoned in many areas with a consequent deterioration of the local habitat. To restore such areas to their former equilibrium, and to help

safeguard their dwindling wildlife, the old cycles need to be re-introduced. Traditionally woodland was harvested for a number of products by coppicing – the regular cutting of underwood trees (ie those growing under the larger timber trees) close to the roots. Many trees including hazel, ash, maple, willow, alder, small leafed lime, chestnut and oak, will not die if they are cut when small in this way, but will produce shoots from the stump (known as a stool or mock) which grow rapidly into long straight poles. If these are cut on a regular basis, in a cycle ranging from seven to fifteen years depending on local tradition and product required, a sustainable crop is produced. The stools themselves can survive regular cropping for centuries. Some are estimated to be over 2000 years old.

When woodland is regularly coppiced it supports a huge variety of wildlife, as there are coups or cants (areas of coppice cut in a season) at all stages of regrowth. For example a freshly cut coup, with the woodland floor open again to the sunlight, will be full of wild flowers, providing nectar for countless invertebrates, in turn feeding small birds and mammals which are hunted by birds of prey. In a few years the same coup will be an ➤

In Britain, unfortunately, apart from where stone was used, or in exceptional circumstances where waterlogging has preserved parts of wooden structures, little remains of the earliest buildings. Our understanding of them is based on archaeological fragments and marks in the ground. Postholes show up as patches of darker soil which replaced the timber posts as they rotted. The cross-section of the posthole after excavation can indicate the post's original size and shape (whether round, squared off, or split) and if the hole was packed with stone.

Houses with posts dug into the

ground are characteristic of Neolithic to early medieval times. The technique makes it possible to counteract lateral thrusts, a particular problem of rectangular buildings, but suffers from the drawback that the posts will eventually rot and make the structure unsafe. The development of freestanding timber frames, which could be isolated from ground water on stone blocks and freed from dependence on earth-fast posts, was made possible by windbracing. Oblique timbers were fixed between vertical posts and horizontal wall plates or tie beams.

➤ impenetrable tangle of undergrowth providing a perfect nesting place for many smaller birds, which may include the nightingale once common but now increasingly rare. Nearby another freshly cut coup will offer rich feeding.

Coppicing is labour intensive. Whereas the woods were once full of human activity now very often all the necessary work is done by a warden or countryside ranger, with the help of volunteers. Where the wood now is mainly cut to conserve wildlife habitats, the poles produced originally supported a wide range of crafts and provided wood for wattle walling and fencing, hurdles, gates, turnery, charcoal and faggots.

By contacting your local outdoor education centre, country park or nature reserve you may be able to organise a visit, between November and March when coppicing takes place, to arrange for your pupils to participate in this valuable conservation work. Ideally, they can learn directly about the practicalities of woodland management, discover the complex and subtle interdependence of species, and produce sufficient wood for an ancient technology project at school. Even a class of 30 Year 5 pupils can, in two or three hours, do a considerable amount of useful coppicing work, given

proper instruction and sufficient adult supervision.

The traditional coppicing tool, examples of which are known from the Iron Age, is the billhook. This is not suitable for use with a large class of young pupils. They can, however, safely use small bow-saws and loppers, although this work should not be undertaken without expert advice.

In a similar way, pupils can learn how other natural products were harvested and in so doing help with conservation work and obtain materials for their own ancient technology projects. For example, thatching reed was traditionally cut in estuaries in various parts of Britain. Just as an overgrown hazel copse progressively declines as a habitat for birds, neglected reed beds also need re-cutting to maintain feeding and nesting sites for sedge and reed warblers and bearded tits. Unless this is done, annual growth declines; the old and fallen stems slow the current and silt is deposited. In a few years the reed bed becomes dry land.

Heather, rush, sedge, bracken and long grass are alternative thatching materials and it may be possible with help to cut some for your own use. The beauty of all these materials is that you are protecting, rather than harming, the habitat if you harvest in the traditional way.

There are other activities involving woodland products that could take place on site or back at school. Heather and thin birch twigs are both used to make besom brooms. A large bundle is tied as tightly as possible with withes or bramble. (The prickles come off easily if pulled firmly through a thickly gloved hand). The handle, which can be of any suitable straight rod, needs one end sharpened to a point, ideally on a shave horse (see page 28). The spike is then driven into the bundle, which tightens it further, by banging the base of the handle firmly on to something hard.

For hundreds of years, until coke was first used for iron smelting in the eighteenth century, the production of charcoal in coppiced woodlands, was a major feature of the landscape. Oak coppice too was widely grown to provide bark for tanning leather. At first cutting down trees to help regenerate a piece of woodland might seem contradictory. But, unless the relationship between the environment and the human beings who have been modifying it since the Ice Age is understood, we have little hope of making sensible decisions about technology and the use of resources in the future.

Piecing together the evidence for ancient houses in Britain gives the following picture. About 6000 years ago (at the start of the Neolithic period) the slow process that changed the principal means of subsistence from hunting and gathering to farming began. Very few remains of dwellings have survived from the long period of time before permanent settlement, presumably because the temporary shelters of the hunter-gatherers were lightly built. In contrast, early farmers started building substantial structures making considerable use of timber except where suitable stone was

loose on the surface. Waterlogged sites in Europe have revealed that these early farmers were already using most of the basic woodworking joints. The later introduction of metal tools made the job of cutting such joints easier but did not radically change the way wood was worked. For reasons that are not clear about 4000 years ago (in the Bronze Age) there was a change in the basic shape of houses in Britain from rectangular to round and this pattern persisted for the next 2000 years until the Romans restored the primacy of the rectangle. Rectangular buildings were nevertheless

constructed throughout the Bronze and Iron Ages in association with roundhouses. They are thought to be raised granaries or store houses partly by analogy with four and six post structures built until recently in Romania and partly from excavation evidence. Strangely, on the continent, where technological developments generally ran parallel to those in Britain, rectangular buildings predominated throughout the prehistoric period, with roundhouses rare but not unknown, except in North West Spain and Brittany, where they were more common.

BUILDING A ROUND HOUSE

Round houses are the simplest, strongest and easiest type of dwelling to construct. Strong winds slide off conical roofs and on hill tops, where many such buildings were erected in the Iron Age, that would have been an important advantage. We do not know how elaborate the original houses were, nor how they were furnished, decorated or fitted out. We do not even know how many people lived in them.

They vary in size but, by about 300 BC, ten metre diameter houses were common and 15-16 metre structures not unusual – no mere mud huts, as a 10 metre building can hold 100 seated adults. It is possible to build such a round house with children (see cover), but such a major project, which would take a full year, is probably beyond the scope of most schools. However, a more modest structure of three to four metres is a realistic project for half a term.

Planning

Enthusiasm can easily run away with time when working on ancient technology projects. Allow plenty and do not be too ambitious to begin with. It makes a better project to base your building as far as possible on local evidence. You may have to scale down the size, but your local archaeological unit or museum should be able to advise you on what has been discovered and where the excavation reports can be found. You will also need to identify sources and availability of the necessary materials. Coppiced wood and reed should be cut in winter to avoid disturbing wildlife. Thought must also be given to where any building will be sited. If possible, choose well drained, level ground away from any modern services and orientate the entrance away from the prevailing wind. Adult helpers may also have to learn or practise a number of essential techniques – such as knot tying. You may also wish to identify one or two local experts – a thatcher perhaps – to help the

Mark out a circle
– mark positions for posts
– make holes with bar
– drive in posts.

Weave wattle – notice the forked posts for the entrance.

Raise the tripod (the first three rafters lashed together)
– use plumb bob to locate apex over centre of circle.

Lash rafters to top of wall
– tie on concentric rings of hazel (400mm spacing).

Tie bundles onto roof using temporary sway
– cut string holding two/three bundles at a time
– push thatch up roof using legatt
– tighten down sway
– use legatt to dress whole roof.

pupils at particular points in the project.

The structure

Apart from variations in the way walls were made, there are two common forms of round house:

■ the rafters rest directly on the walls (the best method for a small school project).

■ the rafters rest on the walls and also on one or more inner rings of beams supported by posts. The latter type of structure was used principally in the largest buildings, although even 10 metre round houses were sometimes built without inner post rings.

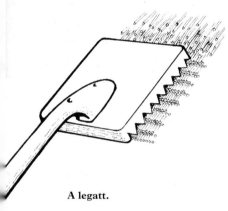

A legatt.

Tools and equipment

■ 2 or 3 legatts for thatching

■ 2 or 3 thatching needles

hazel spikes to hold bundles of thatch onto roof before tying

2 ladders

saw for trimming rafters

■ mallet or post driver

axe and chopping block for sharpening posts.

Materials

The basic rule about prehistoric building materials is that, before the Romans constructed their roads, moving heavy goods long distances was not viable, unless like iron or pots of wine or salt they were very valuable. So local

resources were exploited. The landscape has altered, however, and it may not be so easy now to cut thatching materials or coppice wood in the vicinity of your school.

The ideal shopping list for a 3.5 metre round house would be as follows:

■ 18 rafters 3.5m long 75mm at butt end

■ 36 sharpened posts 1.3m long 75mm at butt end (2 of them thicker and ideally forked to support lintel)

■ 6 substantial (300mm thick) bundles of thin (20-30mm) withes/hazel rods

■ 1 lintel 1.3m (slight curve an advantage – doorway higher in middle)

■ 30 metres of long pliable wood (40mm at butt end) to bend round roof to form concentric rings 350mm apart (hazel is good to use)

■ 30 metres of long pliable wood (30mm at butt end) to form sways for tying down the thatch (hazel is good to use)

■ 100 (300mm diameter) bundles of water reed (or equivalent) for thatching

■ 1.5 tonnes of clean clay or clayey soil soaking in dustbins or old bath

■ 0.25 tonnes cow dung (optional)

■ 1 horsehair mattress and/or two bales of hay or straw

■ 200m strong thick (3-4mm) string to lash building and tie on thatch.

Each type of tree produces wood with particular properties. For wattle walls hazel or willow will weave most easily, but stiffer woods can be accommodated by widening the gaps between the posts. Ash is ideal for rafters, but non native

species like sycamore are also self seeding and often grow close, tall and straight. Green wood (ie freshly cut with the sap still in it) is softer and more pliable than seasoned timber and is appropriate for most ancient technology projects. Thatching is easiest with water reed, because of the length of its stems, but straw, hay, grass, rush, bracken, heather or seaweed could be used.

Walls can be made from flat stone (dry stone walling or with clay or mud mortar), turf (in staggered layers like bricklaying), vertical wooden stakes (with gaps plugged with mud), cob (see page 3), or wattle and daub (this method is most fun and easiest). The height of the walls is rarely evident on excavated sites so your pupils should decide the appropriate dimensions for the entrance and where the roof meets the walls.

To prevent the rafters from pushing the walls outwards (not really a problem on a small structure) they are best lashed together using concentric rings of pliable wood such as hazel. These then provide convenient fixing points for attaching the thatch. The rafters need to be pitched at about 45 degrees; too shallow – the roof will leak, too steep – too much thatch will be needed. Many round houses seem to have been equipped with porches facing away from the prevailing wind. It would be complicated to add a porch to a 3.5 metre hut, but the thatch could be extended over the entrance to keep rain from the threshold.

Methods

There are no hard and fast rules and through guided discussion a class should be able to work out an appropriate sequence and satisfactory techniques for most of the work. What your pupils may not be practised in is working co-operatively on physical tasks and containing their enthusiasm for what they will probably see as 'real' work. It is also important that adequate knot tying is taught because the building proposed here is held together by its lashings. No evidence has survived to tell us

Testing a coracle – sitting in the middle of the coracle is easier!

how roofs were fixed together. There are various possible solutions. Lashings are recommended for safety and strength. Thatching is quite straight forward and involves fixing successive layers on top of one another tightly tied down by means of a sway (bendy hazel rod) tied to the inner concentric rings by means of a long wooden needle. A legatt is a traditional tool for batting the ends of the reeds to an even finish and until tight. It should not be overdone – a real possibility because everyone loves to bash away with it! No-one knows exactly how prehistoric thatch was fixed to the roof, nor how thickly it was laid. Whether a roof lasted 5 or 50 years depended on the quality of the thatching as did the life of the building itself.

The Sweet Track, 3806-3791 BC. An example of a timber trackway constructed over 5,000 years ago and recently excavated by archaeologists.

OTHER STRUCTURES

Flexible structures, made by weaving freshly cut – ie green – lengths of hazel or willow (withes), are very strong and versatile and have been widely used from the earliest times. The only essential tool needed is a blade to cut the stems. Baskets and traps were made in this way, as were the family of coracle type boats. Domed, hide-covered houses, like larger inverted coracles, were probably an important feature of nomadic life. Wattle, or woven,

fences, like the walls of the roundhouse described above, must have become important once farming began. Tipped over, wattle work made fine trackways for crossing boggy ground.

Heavier timbers – split to make planks – were also used for trackways and if there is boggy ground or a ditch that needs bridging this could be attempted with pupils. Splitting can be accomplished by quite young pupils using wedges, wooden or metal, if care is taken to debark the

timber (oak or sweet chestnut are ideal) and steer the split back on course if it deviates. The grain of the log chosen must be more or less straight to achieve a satisfactory result.

Simple rectangular buildings with rigid posts can be constructed to keep the rain off an outside oven or kiln or as shelter for activities. The secret of packing a post so it cannot move is to backfill the ground 50mm at a time and pound it hard before adding more soil or stone.

FIRE ACTIVITIES

Fire, like earth, is something with which young children have little direct experience of these days. Fire which provided heat, light, protection and cooked food, and later enabled pottery, metals and glass to be produced, must have been at the heart of human activity from the start. Until quite recently fire was the major source of energy available for all forms of industry. Fire could be dangerous, and babies were presumably taught to respect it as we teach our young to have due regard for electricity and traffic now. But fire was much more than just a useful tool.

It is not clear when people first learned to make fire, as opposed to capturing it from natural conflagrations, but there seems to be an association between having hearths, sharing food and developing language. Certainly, at least by 700,000 years ago as they colonised temperate latitudes, early humans seem to have grown increasingly dependent on its use. The fire burning in the hearth at the centre of the home, its flames peopling the stories that were probably the first instruments of learning, seems to have a universal attraction. The fascination is still there and this section looks at how this interest can be channelled constructively. Start by simply getting your pupils to discuss the benefits of fire.

MAKING FIRE

Every pupil seems to know that it is possible to make fire by rubbing sticks together. It is, but it is not quite as easy as it sounds. Some practice is required. But both this method, which has left no archaeological traces, and the percussive flint technique which has, can be mastered and used successfully by Year 4 pupils onwards.

The two sticks – or fire bow – method of the two is perhaps better suited to younger children. Skill and a great deal of effort are

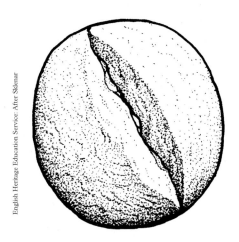

rewarded first by a growing spiral of smoke and then by the glowing red ember which signifies success. You should not take it for granted that even the most elementary facts, such as that heat rises, are automatically understood. There is

Striking a sharp flint against iron pyrites can make sparks which can be caught and fanned into an ember for fire making. Certain dried fungi and kinds of punk (rotten wood) are suitable for keeping a spark alive. This worn piece of iron pyrite (c. 70mm diameter) is over 15,000 years old.

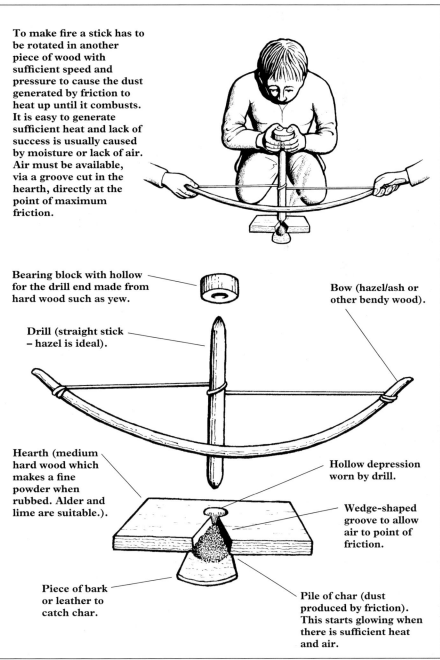

To make fire a stick has to be rotated in another piece of wood with sufficient speed and pressure to cause the dust generated by friction to heat up until it combusts. It is easy to generate sufficient heat and lack of success is usually caused by moisture or lack of air. Air must be available, via a groove cut in the hearth, directly at the point of maximum friction.

Bearing block with hollow for the drill end made from hard wood such as yew.

Bow (hazel/ash or other bendy wood).

Drill (straight stick – hazel is ideal).

Hearth (medium hard wood which makes a fine powder when rubbed. Alder and lime are suitable.).

Hollow depression worn by drill.

Wedge-shaped groove to allow air to point of friction.

Piece of bark or leather to catch char.

Pile of char (dust produced by friction). This starts glowing when there is sufficient heat and air.

English Heritage Education Service: After Sklenar

English Heritage Education Service

much to learn and teach about basic materials and processes in making and building a fire. You will never teach a more rewarding lesson if you succeed – so practice your own technique first! Once completed you are in the perfect position to discuss how Western science teaches the world itself was formed out of a ball of fire. The activity can be developed by telling origin myths from around the world that are based upon fire.

LIGHT

The ability to make artificial light gave early humans a huge advantage over other animals. Dark caves could be entered and inhabited and predators such as bears forced out. Fire-light and then lamp-light extended the day making it possible for many of the activities mentioned in these pages to continue into the hours of darkness. Later on it was stone lamps filled with animal fat and grass wicks that lit the deep caves where drawings and engravings were made and paintings daubed with pigments crushed from rock.

Stone and rush lamps (use lard as fuel) and pigment making (either water or lanolin as binder) are ideal activities particularly for younger children, especially in summer when they can apply the paints to themselves.

COOKING

Apart from roasting food on a spit, or cooking in the ashes there are a number of 'ancient' methods that can cook food very effectively. Various native and wild plants which are common, but no longer regularly used in cooking, can add flavour. Unfortunately, modern scruples concerning hygiene may mean that experiments using these methods have to remain untasted by all but the teacher!

■ Boiling in tub or pit lined with wood and sealed with clay; by dropping in stones heated on a bonfire. **Warning:** some rocks explode in an open fire. Loading and removing stones should be done by an adult, appropriately protected, while pupils watch from

a minimum of 5 metres.

■ Cooking in clay – no foil needed; best done in clay oven.

■ Smoking – to preserve food; traditionally done by building a box designed to funnel smoke from a fire of oak or applewood chippings onto the fish or ham.

■ Cooking in leaves under the embers of a fire (use dockleaves).

■ Baking – bread can be baked in a clay oven very successfully – and the oven built easily.

CLAY STRUCTURES

Ovens, kilns, furnaces and smithing hearths can be grouped together as they share the same basic building techniques and materials and rely upon fire to function. Each can be constructed

Jake Keen

Feeding the oven

over a wattle framework of bent hazel or willow covered with grass, or built up in 300mm stages like a huge inverted coil pot, allowing a day between each stage to allow the previous layer to harden up. Dig the clay from your nearest natural source (it is very expensive to use clay bought for pottery). Add as much grog (crushed up brick, tile or old pottery) or temper (sand, grit, or finely crushed shell or charcoal, hair or chopped plant fibres) as it will hold without falling apart (usually about 30% – a silty or sandy clay will need less, a ball clay more). The grog and temper help the clay to dry out quickly and

thoroughly and limit the cracking which will occur when heat is applied. If the clay is dry enough to crush to powder mixing in the grog or temper will be much easier. Add water little by little until the mixture becomes workable.

It is an advantage to build these structures with thick walls (200mm). They are stronger this way, and being better insulated, are cooler on the outside – and therefore safer to touch – and less fuel is needed. As with clay pots it is good to allow as much time for drying out as possible before firing, although some very silty clays can survive immediate firing.

The dome shaped oven illustrated here has no flue or chimney, but the wood fire within stays lit because air enters and smoke leaves through the same arched opening. The fire (use well seasoned wood) takes up to three

and a half hours to heat the clay walls sufficiently to bake bread. Either let the fire burn down or remove the embers (you can bury them to make charcoal) before putting in the uncooked loaves. Block the opening with a wooden or clay door to keep in the heat.

Ovens of this kind, or kilns which have a similar form or the furnaces as illustrated, are capable of containing temperatures of well over 1000° centigrade while the outside stays only warm. Sensible precautions should be taken to ensure pupils do not wander freely near the opening while the fire is

Preparing a clamp firing.

lit. Enhanced temperatures can be achieved by using bellows, which can be of conventional design, or the result of a design and technology project (there are various historical solutions). If recording the temperatures is to be part of Science investigations thermocouples should be installed by poking them through the wet clay while the structure is being built.

Pottery

Heavy, fragile clay pots are of little use to nomadic people who usually prefer to carry lighter artefacts, but once farming and permanent settlements became established, pottery with its great versatility and scope for decoration appears in the archaeological record. Because styles of pottery are easily identifiable and tended to be consistent over long periods, ceramics are very useful for dating and have been much studied by archaeologists.

Most pupils enjoy making and decorating pots, but the exciting stage at which dry brittle clay is transformed into something akin to stone, usually takes place out of sight and reach in the steel box of an electric kiln. Pottery can be fired successfully and safely with children – in such a way that they can observe and participate in the process – in a variety of ancient ways. In all of these methods it helps if the clay has been well tempered (see above) with

appropriate material (local museums may be able to lend you typical examples of early pottery from your area to study). This should be done, if you are using freshly dug clay, after removing worms, stones, roots and any other debris.

There are three main methods of firing:

- Clamp or pit firing

- Bonfire firing

- Kiln firing.

The crucial thing to remember is that if moisture is trapped inside the thick wall of a pot it will expand and break the vessel as it heats up and turns to steam. To avoid such breakages make sure:

- to temper the pots well, working the clay to remove air pockets

- to allow the pots to dry well before firing

- the initial rise in temperature from 0 – 200° centigrade takes place slowly and steadily.

Experimental Romano-British style kiln with modern flue.

A bonfire firing in progress.

Retrieving pots after a clamp firing – always a popular activity!

FOOD AND FARMING

The conventional view that the lives of our ancestors were dominated by the incessant search for food until rescued from insecurity and hunger by farming – which in turn led to civilization is now seen as false.

Recent evidence has shown that hunter gatherers living today in the 'marginal' lands left to them spend less time working for their food, have a better diet and stay healthier than the average subsistence farmer. Some archaeological evidence suggests that the adoption of agriculture led at first to a significant reduction in skeleton size, indicating poor nutrition, and a great increase in infectious diseases caused by germs which could survive only in crowded permanent settlements.

Farming can produce more food per hectare than hunting and gathering and is therefore capable of supporting a greater population. Some of the long term developments usually associated with farming are listed below and can provide a framework for discussing the activities that follow. Not all developments in the list are necessarily restricted to farming communities – for example most hunter-gatherer societies probably functioned within well-defined boundaries and had well-established and complex ceremonies – and pupils should be encouraged to discuss the list and how different elements in it would have changed society.

Changes associated with the introduction of farming:

■ increased need to plan ahead – prepare the ground and sow for a harvest months ahead.

■ permanent settlement and associated services (for example, water supply).

■ increased clearance of land led in places to deforestation and

Year 8 pupils carrying out experimental ploughing with a replica plough.

major changes to the natural habitat.

■ selective breeding of animals and plants.

■ surplus and storage of food; wealth creation.

■ specialists producing prestige goods.

■ class divisions: hierarchies; elite of non producers; artisans, workers, slaves.

■ development of boundaries, politics, armies, warfare.

■ development of organised labour, administration, administrative and ceremonial structures.

■ increased trade, improved communications, city states, nation states.

■ expansion, colonisation, commerce, industrialisation.

The activities suggested below are designed to give pupils insight into how, little by little, through such developments society has been transformed.

AGRICULTURE

There is evidence for farming in Britain from about 6000 years ago.

Marks scored in subsoil or chalk indicate the use of simple ploughs or ards. These can be made quite easily and pupils instead of oxen used to pull them along.

Burnt seeds are frequent finds on archaeological sites and indicate which varieties of wheat, barley, peas, beans and wild flowers grew in prehistoric fields. Though in short supply, ancient seed types are sometimes available from agricultural research establishments and valuable work with them can be done in school. If there is a patch of ground available it could be ploughed up with an ard or the surface broken with digging sticks, cleared of weeds, and planted in rows. Different ways of improving the soil could be tried and the results compared by measuring the height of stem or weight of crop for each row tested. Various types of animal dung, woodash, compost, leaves, mown grass, etc. can produce results surprisingly better, or sometimes worse, than the control (unimproved) strip. Another strip could be left unplanted to see what wild seeds have been waiting dormant in the soil. Some seeds can survive several decades under turf and still remain fertile. Before chemical spraying removed the wild flowers, cornfields were full of colour.

Crops of wheat, whether winter or spring sown, could be harvested early in September by the following year's class, who could do germination experiments (100 seeds per tray between wet paper towels) while learning about percentages before planting out their own seeds and renewing the cycle. It may take several seasons to build up sufficient seed stock to allow for bread making, but threshing with a flail, winnowing to separate the chaff from the grain, and grinding with a quernstone can be done with modern wheat. Most pupils enjoy discovering how flour was first ground by rubbing the grain between two flat stones. Saddle querns can be made quite quickly and easily by chiselling out

Working hard at a rotary quern.

a shallow hollow in soft stone (gritty sandstone is the best). To grind enough flour for the daily bread on a saddle quern takes much time and effort (90 minutes per kilo for an adult). An early example of advancing technology is the rotary quern which began to displace its simpler predecessor in Britain in the late Iron Age. A greater investment in time was needed to make it, but an adult using the rotary quern could produce a kilo of flour in about 10 minutes. Animal, water, wind and eventually electrically powered mills continued the development process. It is not certain when bread was first leavened, but the flour produced on the quernstones can be made into small loaves with water alone, or by adding yeast, honey, fruit, milk, cheese or other appropriate ingredients and baked in a clay oven.

ANIMAL HUSBANDRY

Few schools are lucky enough to keep their own livestock, but related work is possible. Making butter and cottage cheese is not difficult. A simple butter-churning lathe is quite easy to construct. A pottery strainer to separate curds from whey could be fired with other pots, loom weights and spindle whorls.

Providing animal feed in winter was always a problem and if a grassy area of the school can be left unmown for a few weeks in early summer haymaking could be attempted. A pet pony or goat might be needed to test the edibility of the results.

FENCING AND EARTHWORKS

Protecting crops from livestock or wild grazing animals requires fences or hedging. Wattle fencing can be undertaken quite easily if you have access to appropriate materials (see page 16). Earthworks too are associated with the first permanent settlements and prehistoric farmsteads were often surrounded by a ditch and bank. Whether or not this is done in association with building a hut,

Concentration, application and successful and tangible outcomes: making a wattle fence with year 4 pupils.

substantial earthworks can be constructed by a class in a short time. Such work can usefully demonstrate the potential of working together and give some measure of the efforts required to build the hill forts, barrows, ditches and banks that have been prominent in the landscape for several millennia.

Modern tools such as steel buckets, shovels and wheelbarrows can be compared for effectiveness with antler picks and mattocks, wooden or bone shovels (shoulder blades of cattle), baskets, leather bags, wooden sledges, or stretchers. Raw materials for some of these unorthodox tools will need hunting out from butchers, deer parks or zoos.

FOOD BEFORE FARMING

Most pupils are intrigued by ingenious inventions for hunting and trapping wild animals. However, making bows and arrows, spears and spear throwers, catapults, blow guns, boomerangs, bolas, rabbit sticks, snares and

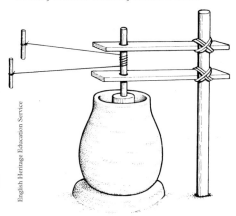

A simple butter-churning lathe.

traps are projects that are difficult to control and can be frustrating as the weapons can rarely be tested in real situations. Nevertheless, some activities can be carried out to give pupils an idea of early subsistence techniques.

■ Flint knapping – (knives, arrow and spear heads) – joining to shaft with beeswax and resin or birch bark glue (see page 26).

■ Bone work – harpoons or fishhooks.

■ Fishing lines and nets (use flax or nettle fibres, or lime or poplar bark).

■ Baskets for collecting and carrying food; fish traps.

■ Wild foods – there are many wild plants which are nutritious and which were utilized in the past (see *Bibliography and resources*).

TOOLS, SIMPLE MACHINES AND ENGINEERING

TOOLS

One of the main things that sets humans apart from other animals is that we make our own tools. Early humans developed a kit of stone implements to cut, saw, scrape, pierce, bore, hammer and chop. Flint was the best material available in Britain, but other stones such as chert, quartz and Langdale tuff were also serviceable. Under strict supervision, pupils can make simple flake tools (blades and scrapers) although much practice is needed to achieve prehistoric standards! As flint shatters like glass, goggles and tough gloves are essential. The resulting tools can then be used to shape bone into needles, toggles, fish-hooks or combs, or to work wood or leather.

The discovery of metals considerably extended the versatility of the tool kit and contemporary illustrations of Roman workshops show that most of the hand tools for wood used today were known at least 2000 years ago. The superiority of iron is often stated, but even copper, which was first regularly used 7000 years ago, could be cold hammered to produce a very good edge.

Copper tools

Natural copper is quite soft (with a measure of about 85 on the Brinell scale of hardness). However, when cold-hammered it can reach 140 on the scale – equivalent to the much harder wrought iron. Some copper ores mined in Oman about 5000 years ago contained nickel, tin and other metals in small amounts. When these ores were smelted and then cast into chisels or axes they had a hardness of about 150 which by skilful hammering could be improved

Chris Wakefield

Testing a copper-pipe chisel.

to 275 – similar to the steel used for modern tools.

To demonstrate that copper can be cold-hammered to produce a useful cutting edge collect some scrap copper pipe. It should be clean, dry and free of corrosion. At the simplest level a 150mm length of 20mm pipe can be flattened at one end to a sharp edge by gently hammering between two pebbles. The edge can then be honed by rubbing on a flat surface with sand and finally tested for effectiveness as a chisel using a stick poked into the open end as a handle.

To be more ambitious scrap copper could be heated up in a crucible in a bed of charcoal and then cast into axe head or chisel moulds prepared by the pupils. The moulds can be made in three ways:

■ Chisel out the shape of the tool

in soft stone or cuttlefish shell.

■ Make a wooden version of the axe/chisel. (Must be simple straight-sided, unflared, wedge shape; cover with well-tempered clay; remove wood; fire clay mould).

■ Make a wax version of the tool. (Can be more complicated shape; cover with clay; fire clay mould).

The crucial rule about pouring molten metal is that the moulds must be entirely free of moisture. Tackle this only with expert help and guidance and with pupils well clear. Once cold, the copper tools need to be gently hammered to a hard shiny finish with pebbles and honed to a sharp edge.

SIMPLE MACHINES

The is no way of telling whether the first inventions came about through trial and error or conceptualisation, but it seems that most of the early civilizations recognised a number of simple machines ie devices which could make a job easier to do. Some could magnify movement, other magnify the effectiveness of a force or change its direction. Most of the mechanisms that make up the complex machines we recognise as modern have been known for thousands of years. The idea that a machine makes a job easier by allowing a small force, the effort, to overcome a large force, the load, is best introduced to pupils by getting them to use the simple machines below and then to attempt the same task without their help.

The inclined plane

Getting to the top of a steep hill or raising a heavy weight is made

The inclined plane.

The wedge.

easier by lengthening the route or using a ramp – this decreases the effort needed but increases the distance to be moved.

The wedge

Like an inclined plane on its edge, the wedge magnifies force. The longer the sides in relation to the width the greater the force applied to split the block of wood.

The lever

Levers can magnify force, like a wedge, or movement. In the drawing, the greater length 'a' in relation to length 'b' the greater the load that the same effort will raise. Archimedes said that if he had a lever long enough he could lift the world. Many throwing devices utilise the fact that for a given load a missile is catapulted further the greater the distance from the fulcrum it is released.

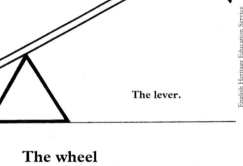

The lever.

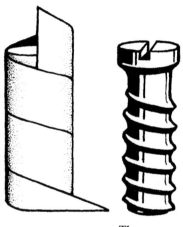

The screw.

The screw

The screw is an inclined plane wrapped round a cylinder. It produces a mechanical advantage because it moves forward with a greater force than is used to turn it. The Archimedes Screw was used to raise water in ancient irrigation systems.

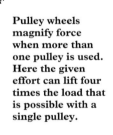

Pulley wheels magnify force when more than one pulley is used. Here the given effort can lift four times the load that is possible with a single pulley.

The wheel

The wheel can be considered as a lever able to rotate 360° round a fulcrum which is the axle. A wheel and axle magnify force because the wheel is larger than the axle.

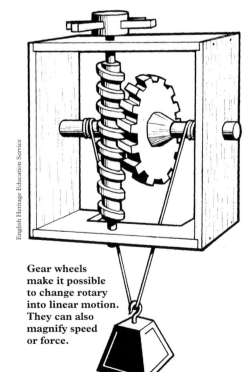

The Archimedes screw.

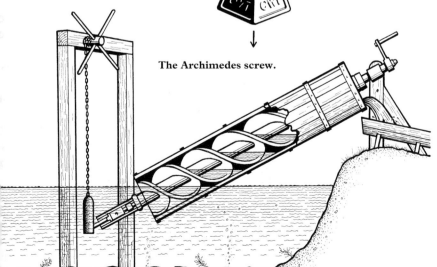

Gear wheels make it possible to change rotary into linear motion. They can also magnify speed or force.

HOLDING DEVICES

Until the carpenters bench with a screw vice was invented in late medieval times other devices, known as brakes, dogs, horses, stools or vices were used to clamp or support wood – freeing both hands to use appropriate tools to work the wood. These elegant inventions are easy to make and are very effective.

The shave horse or foot-vice, just like the milking stool, has three legs so it will be stable on any ground. The vice is closed on the work piece by pushing forward on the footrests. The wedge under the work surface can be adjusted to accommodate wood of different sizes. Pupils enjoy using the shave horse to debark and shape green wood poles to make handles, tent pegs, thatching needles and the like. The tool usually used with a shave horse is the drawknife, which looks dangerous because of its long exposed blade, but is actually very safe if treated with respect, because it has two handles. Alternatively, the more modern (enclosed blade) spokeshave can be used.

The knee vice and tension brake are other ways of gripping a piece of work in progress.

A pole lathe.

Splitting, cleaving or riving brakes enable a pole such as hazel, willow, ash or sweet chestnut to be dividedinto two equal parts by leverage. Riving as a method of reducing a length of wood to thinner pieces is thousands of years older than sawing, is much quicker and produces a stronger result as the fibres are not cut. The finished product is ideal for hurdle making and wattle fencing. The traditional tool to use is the froe

A huge stone monolith is being moved on rollers by a tribal group in Indonesia.

but a blunt billhook is fine for small wood.

The pole lathe, known in this country for at least two thousand years, is a safe and simple machine which utilizes a springy pole attached to a treadle to rotate wood back and forth against a hand held chisel. Bobbins, spindles, rolling pins, chair legs, mallets, bowls and bracelets can be 'turned' in this way.

A shadoof – one o the oldest known machines.

ENGINEERING

In about 4000 BC in Sumer (modern Iraq) the first known city states grew up between the Tigris and Euphrates rivers, dependent upon irrigation. More than 100,000 kilometres of water conduits were installed to enable dry land to be cultivated and to supply the centres of population. To channel water long distances requires knowledge of surveying and some basic instruments. A Roman device known as a groma can be made and used by pupils to survey a stretch of land. The groma was an instrument to set out straight lines. It consisted of a vertical pole with a right angle extension at its top. Four plumb bobs hung from the ends of the arms of a cross that was pivoted on the extension. An assistant would hold a second pole some distance away while the surveyor lined up the second pole with two of the plumb bobs and the main pole of the groma.

If you have the chance to lay a conduit (plastic guttering could be used instead of the authentic stone or ceramic drainage pipe) to enable the groma to be purposefully employed, a wealth of opportunities for using the moving water could be created. Work on model water mills (horizontal, undershot, overshot) and other water powered machines, locks and boat design becomes possible. The water supply could perhaps incorporate a shadoof, one of the oldest and most elegant mechanisms known, and still used in Egypt.

The irrigation works of Sumer, though staggering in scale, are overshadowed by other feats of engineering accomplished in prehistoric times. The Egyptian pyramids are extraordinary achievements for any period. The size and number of blocks of stone, the precision with which they were fitted and the accuracy of their alignments with the planets are difficult to comprehend when it is realised all this was managed without the powerful machinery and scientific knowledge we take for granted today. Avebury and

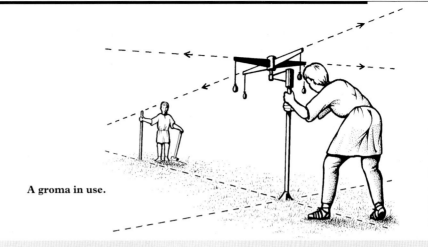

A groma in use.

Green Woodworking

The use of seasoned timber is a relatively recent innovation and for tens of thousands of years most wood was probably worked straight from the tree – while it was soft and could be split and worked easily. Even oak cuts like cheese when green. There is a sweet smell when shavings of freshly cut wood are removed with a draw knife, each timber having its own scent. As the finished product dries it shrinks and may distort. Furniture made from riven green wood is stronger, weight for weight, than that from seasoned timber, but is less symmetrical.

Treen is the old Saxon word for domestic utensils made from green wood and a variety of possible items can be made by pupils. Simple spoons can be whittled and small bowls hollowed quite easily and quickly.

Stonehenge, though dwarfed by the pyramids are, nevertheless, massive achievements, which inform us that considerable social organisation must have existed in Britain over 5000 years ago.

Seventy Year 5 pupils pull a 5.5 ton oak log.

Moving a heavy load by means of rollers and levers

A tree trunk, concrete lamp-post, or other substantial monolith is required. A pair of levers and suitable blocks to act as fulcra can be used to raise the load to insert the rollers one by one. Take great care to ensure that the levers cannot slip before positioning the rollers. The smallest pupil in the group, given a long enough lever and appropriately placed fulcrum, can be granted the satisfaction of

A scene from a wall painting in an Egyptian tomb showing a bow drill in use.

lifting the weight alone. A class of young children can move large logs quite easily if the placing of rollers is supervised closely.

Rotation

It is often said that the wheel has been the greatest human invention. Certainly the idea of rotation is at the heart of many of the most useful early labour-saving devices. The 'bowanarrow' is a word used by most young children to describe a curved stick with a taut string tied to each end. In fact, apart from propelling missiles, the bow was used from the earliest times to convert backwards-forwards reciprocating motion of the hand into clockwise-anticlockwise rotary motion of various tools such as the drill, lathe and fire maker.

SOFT TECHNOLOGY

In the list of human needs clothing usually comes after food and shelter. In all but a few parts of the world people wear clothes to go about their daily business. It is assumed that the earliest inhabitants of northern latitudes dressed in the skins and furs of the animals they hunted. Nomadic hunting peoples for whom records exist, give us a consistent picture of the full use to which their prey was put. Hide, horn, hoof, sinew, bone, flesh, brain, intestines – nothing was wasted.

Farming and the domestication of animals reorganised the exploitation of animal products which became the raw material for a number of specialist trades. For example, leather working occupied up to 25% of the tax paying population of many British towns in the medieval period. For many hundreds of years leather manufacturing was second only in economic importance to the woollen industry.

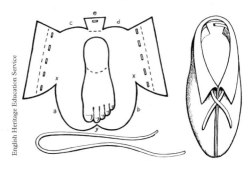

Mark outline of foot with x-x being the circumference of instep. Turn over. Stitch edges a to b, starting at x finishing at y. Stitch c to d. Fold e inside. Turn moccasin inside out. Thong slips under the flaps. (Source: Coles 1979).

The Ice Man wore leather shoes, each made from an oval of cow-hide with the edges turned up and bound with leather straps. The straps were attached to a net which held in place the grass which kept his feet warm. A simpler form of footwear, easily made from scrap leather, is the Woodland Indian moccasin. A more complicated pattern, but also using one piece of

hide, was adopted by the Romans. Watertight containers, made by folding a sheet of flexible material, such as birch bark or a strong leaf, are still used in many parts of the world. The technique can be practised using a sheet of paper and a couple of pins. The Ice Man carried two birch bark containers but these were sewn rather than pinned.

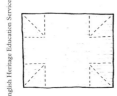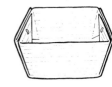

Leaf or birch bark container (source: Coles 1979).

Fibres from both plants and animals must have been used for tying, lashing and sewing from the earliest times. The oldest surviving example of woven fibres come from Çatal Hüyük in Turkey and are about 8500 years old. In Britain, spindle whorls (perforated disks of baked clay or stone, which acted as flywheels for the spindles), bone weaving combs, and loom weights (again of baked clay or stone) appear in the archaeological record from about five thousand years ago. Such finds are common on European Bronze and Iron Age sites and, although no looms as such have been found, evidence in the form of depictions on Greek vases, postholes associated with lines of loom weights, and woven clothing preserved in peat bogs, gives us a fair idea of how weaving was done.

Making a full scale working warp-weighted loom is a viable project but it can take several hours to set up and only a few minutes to get hopelessly tangled. Several looms – perhaps one to a group of four children – would be manageable, but an adult volunteer with both patience and expertise to assist in tying the correct warp threads to the appropriate heddle rod would be necessary. Small weaving frames

can be made by individual pupils that will at least give them an idea of the process of weaving.

Weaving requires spun yarn but spinning too, though simple to understand, takes patience to master and can be very frustrating. It is the ideal kind of activity to practice at odd moments (perhaps breaktimes) when the pressure is off. Even the best wool has a staple (length of fibre) of only 100mm or so and an easier introduction to spinning is made possible by using either of two plants exploited in prehistory, flax and nettle, which have much longer fibres. These can be prepared quite easily from the plant stems. Cut the plants before the seeds mature and remove all leaves. The stems then need to be retted (partly allowed to rot). Do this by placing them in water for a few days – a very smelly process. The retted fibres can then be spun and woven. You can supplement your supply of retted fibres by mail order (see *Bibliography and resources*).

The fibrous stems, leaves, seedpods and barks of different plants can be used to produce string and rope. In Britain, apart from flax and nettle, the inner bark of lime, poplar and elm produces strong fibres. These can be loosened and separated in twenty minutes or so by gently pounding. A better result is obtained by retting for six weeks – especially if the fibres can be retted in the soft mud at the edge of a river or stream. The astoundingly smelly fibres can then be washed and either twisted or plaited into string or rope – as happened at Bembridge school (see page 14). Lengths of fibres, loose, twisted or plaited, can be coiled to make mats or baskets, each coil being joined to its neighbour by sewing or wrapping.

Easier to master than either spinning or weaving, but of equal antiquity is sprang, which is a technique of twisting strands

English Heritage Education Service

of wool together, very much like the children's game of cats cradle. It is not weaving or knitting as no extra threads are added during the process of making it. Sprang hairnets and bonnets have been found preserved in several prehistoric graves in Denmark. The elastic quality of the sprang material makes it ideal for fashioning bags or nets to contain moss for pillows or cushions – a much more economic use of wool than woven cloth.

Making a sprang hat or bag

Make the frame as shown and wind wool round the strings at least 15 times **❶**. Starting from left or right twist the front strand round the back of the next one maintaining the twist by inserting a stick between the twisted strands as you work across the frame **❷**. Begin the next row in the same way twisting the front strand behind the back and inserting a second stick to maintain the twist. After each row of twists take out

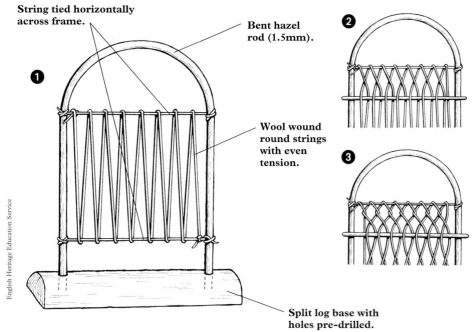

String tied horizontally across frame.

Bent hazel rod (1.5mm).

Wool wound round strings with even tension.

Split log base with holes pre-drilled.

English Heritage Education Service

the penultimate stick ready for the next row **❸**.

You will see the lattice-like pattern developing. You will also notice that for each row twisted at the top of the frame another will have formed at the bottom and two

sticks will be needed here as well to prevent unravelling. As you reach the middle, instead of a stick, loop wool between the twisted strands. Now remove the sprang net from the frame and sew up the sides to make a cap or bag.

The Ice Man

In 1991 a frozen body was discovered emerging from a glacier in the Alps. Tests eventually determined that the almost perfectly preserved, clothed body was that of a man who died over 5000 years ago.

Usually artefacts from prehistory only reveal a partial picture. Tools lack their wooden handles, leather, string and other soft materials have almost invariably disappeared and there is no way of knowing if the objects buried in graves are special or commonplace. Here, in a single chance find, a wealth of new

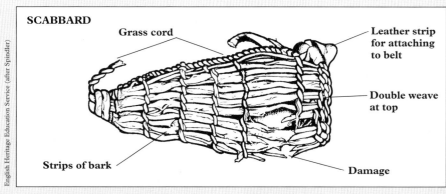

SCABBARD

Grass cord

Leather strip for attaching to belt

Double weave at top

Strips of bark

Damage

English Heritage Education Service (after Spindler)

evidence about ancient technology has been revealed. Apart from carrying a copper axe (see page 26) the Ice Man possessed several other items which may be copied

and tried out.

Retoucher Looking like the stub of a fat pencil this tool was probably for finishing the edges of flint tools such as this knife.

Scabbard To make the Ice Man's knife's scabbard tie one end of 16 lengths of bast (or grass or string) together. Weave two strands over and under 7 times. Fold the result in half and sew up the long side.

Back-pack frame Made from a two metre rod of hazel bent into a 'U'. The two larch wood boards were notched at each end and lashed to corresponding notches in the hazel rod.

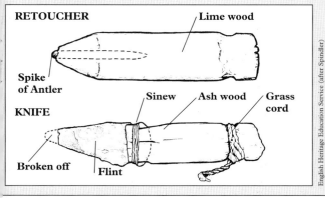

RETOUCHER

Lime wood

Spike of Antler

KNIFE

Sinew   Ash wood   Grass cord

Broken off   Flint

BACK PACK FRAME

English Heritage Education Service (after Spindler)

ANCIENT TECHNOLOGY ACROSS THE CURRICULUM

Ancient technology can provide a framework or stimulus for work across the curriculum. While most of the ideas suggested on the following pages are listed under specific subject areas, the majority are cross-curricular.

HISTORY

On the surface, the history National Curriculum seems to exclude the study of prehistory – from where many of the examples above are drawn. However, prehistory is, at the very least, an essential introduction to Roman Britain and can be used to great success as the basis for a local study or study of an era or major turning point in the past. By having close, hands-on experience of some of the technology of prehistory pupils can begin to understand something of what it may have been like in the distant past – and through such investigation – begin to look more closely and critically at their own society. A number of technological devices, such as the shadoof, were important developments in the rise of other societies studied under National Curriculum guidelines.

GEOGRAPHY

Schools looking at 'developing' countries, or exploring alternative materials and sources of power, may find relevant starting points in ancient technology. Some development programmes specifically examine how to improve an ancient technique to conserve materials or to save time. For example, clay ovens, which have been used for thousands of years, are still found in every home in parts of the world. Many are inefficient and waste wood or

A roundhouse is an ideal venue for storytelling.

charcoal, a serious problem where deforestation has occurred. Designs for economic models can be developed and tested by pupils.

Appropriate technology

In the past aid from industrialised to undeveloped countries has sometimes involved transplanting inappropriate technology, hardware or machinery that cannot be maintained or serviced without a highly evolved infrastructure which the poorer country frequently lacks. Even in the remotest rural settlements where traditional technology continues to shape the way of life modern inventions co-exist. For example, often an excessive proportion of a family's budget is spent on batteries to power torches and radios. A wind-up (clockwork) radio has recently been invented to get around this problem. What other needs of developing countries can your pupils think of? Can they work out an appropriate intermediate technology solution?

We tend to take the supply of clean water for granted in Britain and squander it without thought. But when water was raised and carried from the well, as still happens in many parts of the world, the yoke and pail were often

used. Many local museums have examples which can be copied easily and used to try out some of the daily tasks of the past. Trying to estimate how time was spent in the past gives insight into how and why change has occurred, and helps us view sympathetically the situation in countries seeking technological development of their own.

ENGLISH

Ancient technology – and especially some of the outcomes of ancient technology projects such as roundhouses – provide excellent opportunities, and environments, for role play, storytelling and various forms of creative writing.

One activity that has proved very successful as a culmination to a project on ancient technology has been where different classes that have been involved in the project have staged a market – occasionally at a relevant historic site. (This has also worked well with pupils from different schools taking part.)

All pupils should bring a few items they have made during the project. Split the pupils into small groups with each group sharing a cloth – to act as a market stall. Objects could include craft items made in school (fired pottery, jewellery, woven materials or cooked and preserved food), fruit in season, 'specialities' of each school, prized natural finds – fossils, unusual stones, or 'imported' goods. Trading should start at a given signal. If the pupils have put effort into what they have made and brought along objects they value themselves, they are likely to wish to acquire items they see on other stalls. The beauty of bartering is that it only works whe

Transport and Trade

From the earliest times there is evidence of the movement of raw materials – flint, amber, ores and whetstones for instance are often found far from their source of origin. It was once assumed that each major development in technology in prehistoric Britain represented a new wave of invaders from the continent. It seems more likely now that innovations, which often occurred in the heart of Europe first, were brought to these islands through long established trading connections. We know from documentary sources, for example, that long before Britain became assimilated into it, there was considerable trade with the Roman Empire. The underwater archaeology of wrecks has revealed a great deal about the sophistication of the ships used, and the great variety of cargo carried. Within the British Isles sewn plank and log boats from the Bronze Age onwards have been found in the mud of estuaries and river channels and, somewhat later, coracles are described by both Caesar and Pliny. The latter are quite feasible to make with children and can provide an excellent summer term finale.

Making a coracle

To make a simple oval coracle you will need

■ 40 rods of hazel or willow (20 1.6m x 25mm at the thick end, 20 1.5/2.0m x 15-20mm)

■ canvas painted with

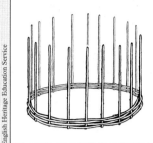

Three stages of coracle building

bituminous paint or plastic sheet (2.0m x 2.4m)

■ 1 plank (1.2m x 150mm x 20mm)

■ a ball of string.

Push the 20 larger rods into the ground in an oval (approx 1.6m x 1.1m). Weave the thinner rods tightly round the base of the vertical sticks to form the gunwale (edge). Bend opposite rods over to meet one another and lash together to make a low, inverted basket shape. Place heavy stones on a plank for a few days to flatten the bottom. Then lever out of the ground, trim off the ends and stretch and sew the covering tightly over the frame. Drill holes at each end of the plank and tie to the gunwale. When trying the boat in the water step carefully into the middle. Propulsion is effected by drawing a long paddle in a figure of eight motion in front of the coracle (see page 20!).

On land, the wheel revolutionized transport, but only where there were animals amenable to domestication and strong enough to pull a wheeled vehicle. In the New World, the wheel was known, but there were no such animals until contact with Europe. Even a simple cart is quite a difficult task for children because wheels are hard to make, but useful work on the evolution of the wheeled vehicle could be attempted by making models to show how wheels developed axles and how they were attached and steered.

Interesting work, too, can be done on the wheelbarrow, varying the size and position of the wheel. The traditional Chinese barrow can be used to carry much heavier loads than the European version.

an agreed deal has been struck and the chances are that complicated triangular trading patterns and brokers and agents tend to emerge unbidden. A possible development to introduce a form of money or token – perhaps even minted at school. The event could be extended by the pupils from different classes or schools telling stories they have created about their work for the period – or a storyteller could explain the importance of the site where the market is taking place, or simply tell an appropriate tale for the period.

ART

An ancient technology project allows great scope for pupils to study the art of the past and also a variety of non-Western cultures. Using these examples as stimuli pupils can create and develop their own art as a response. Two examples follow, many more are appropriate.

Soft sandstone and some limestones can be carved with a flint or any pointed metal object. Carved heads from the Iron Age are found in various parts of Britain and the tradition perhaps continued in churches as gargoyles.

Excellent colours can be obtained by grinding iron ore, clays and

Ancient technology and the environment

It is now generally accepted that the demands of our energy hungry, technologically developed way of life are destroying irrevocably a great deal of the natural world we inhabit. Commentators have been remarking on the accelerating nature of technological change for much of this century. Technology, once the means by which human beings solved their practical problems, now defines the context of such problems as well as their solution.

Knowledge of earlier technologies should give pupils a useful perspective, a long view of history, to enable them to assess and evaluate the costs and the benefits of new technology. Without a sense of where we have come from, it is very difficult to see a destination or to plan for the future, and rapid change becomes bewildering. It is important for pupils growing up in the late twentieth century to understand that we are all citizens of one world. Earlier peoples in Britain shared the same problems of survival with those in the rest of Europe, Asia, Africa, the Americas and Oceania: how to find food, make shelter, clothing, tools, utensils, fire. We now share problems such as overpopulation, global warming and extensive pollution.

Pupils should use their work on ancient technology as a stimulus for a debate on modern technology and the future. If their project has included building a roundhouse or other shelter this could be used as a venue for the debate. The debate could be controlled by a talking stick – with the only person allowed to talk being the one holding the stick. Alternatively, pupils may wish to invite local decision makers – perhaps planners or local politicians – to take part in a debate or to answer questions. If well-briefed, such visitors could be interviewed by a panel of pupils about a particular local concern, for example the use of modern technology in traffic control or the local transport policy as part of national policy, with the rest of the class voting on the outcome.

However the debate is structured pupils need to realise that decisions about technology taken over the next few years will have an enormous impact on their future.

A sandstone head carved by a Year 8 pupil as part of an ancient technology project.

charcoal into fine powder and mixing either with lanolin or some other binder.

MATHEMATICS

Many of the activities mentioned above are rich in mathematical potential. For example, the laying out of a circle as the first stage of building a roundhouse and the equidistant spacing of posts around the circle. Pupils could also create their own form of measure based on body parts.

If your pupils have been carrying out experiments as part of the project – for example, firing pots made from different types of clay and with different tempers, they should be given the chance to classify and sort the data by recording the results. Graphs, charts, pie diagrams or other pictorial analyses should be made to illustrate their findings. These can either be hand-drawn or, where appropriate, pupils should use computers to generate their results.

SCIENCE

Ancient technology is really practical scientific enquiry. Pupils should be encouraged to plan their own investigations, to decide how they are going to collect and consider evidence (taking into account all relevant health and safety issues), and how they are going to communicate their results. They should be given the opportunity to use IT to collect, store, retrieve and present their conclusions using standard measures and SI units where appropriate. A number of investigations – for example making and testing different recipes for glue – will address how materials change while other - for example moving heavy weights with levers – will address the properties of forces and motion.

Pupils experimenting with a large xylophone made from small leaf limewood

MUSIC

The earliest evidence for musical instruments comes from bone whistles and bull roarers, but since all known people have percussion instruments of one form or another, it can be rewarding for pupils to create their own, especially if you intend to add dance and story telling to your project. Claves and wooden xylophones, which can make use of offcuts from building or fencing, are very easy to produce.

BIBLIOGRAPHY AND RESOURCES

* Either written for or usable by younger pupils.

GENERAL ARCHAEOLOGY/ TECHNOLOGY

Abbott, M, **Green Woodwork – Working with Wood the Natural Way,** GMC Publications, 1989, ISBN 0-946819-18-1.

Audouze, F, & Buchsenschutz, O, **Towns, Villages and Countryside of Celtic Europe,** Book Club Associates, 1992. Excellent, particularly on structures.

Bahn, P, & Renfrew, C, **Archaeology: Theories, Methods and Practice,** Thames & Hudson 1991. ISBN 0-500-27605-6- Readable, scholarly, comprehensive and up to date.

Burstall, AF, **Simple Working Models of Historic Machines,** Arnold Ltd, 1968, ISBN 0-7131-3242-6.

* Caselli, G, **History of Every Day Things,** Beehive, 1993.

Clark, J G D, **Excavations at Star Carr,** Cambridge University Press, 1972, ISBN 0-521-08394-X.

Coles, J, **Experimental Archaeology,** Academic Press, 1979, ISBN 0-12-179752-X. The classic text, fundamental to exploring ancient technology.

* Cork, B, & Reid, S, **The Young Scientist Book of Archaeology,** Osborne, 1984, ISBN 0-86020-865-6.

Diamond, J, **The Rise and Fall of the Third Chimpanzee,** Vintage, 1991, ISBN 1-09-991380-1. Provocative view of human development (including technology) from a biological perspective.

Dorset Outdoor Education Service, **Safety Policy Document,** 1996, Professional Development Services, Dorset County Council.

Duly, C, **The Houses of Mankind,** Thames and Hudson, 1979, ISBN 0-500-6005-3.

Earwood, C, **Domestic Wooden Artifacts,** University of Exeter Press, 1993, ISBN 0-85989-389-8.

* **Eyewitness Guides,** Dorling Kindersley. Superb illustrations, including Hart, G, **Ancient Egypt,** 1991, ISBN 0-86318-444-8, and **Early People,** 1989, ISBN 0-86518-342-5.

Feinberg, W, **Lost-wax casting, A Practitioner's Manual,** Intermediate Technology Publications, 1983, ISBN 0-903031-88-4.

Gabriel, S, & Goymer, S, **Basketry Techniques,** David & Charles, 1991, ISBN 0-71539-424-X.

Goodwin, L, **A Dyer's Manual,** Pelham Books, 1982, ISBN 0-7207-1327-7.

* Green, M, **Roman Technology and Crafts,** Longman, 1979, ISBN 0-582-20162-4.

Gregory, R L, **Mind in Science,** Peregrine Books, 1981, ISBN 0-14-055168-9. Very good on the link between human development and technology.

Hall, N, **Thatching: A Handbook** Intermediate Technology Publications, 1988, ISBN 1-85339-060-7.

Henson, D, **Teaching Archaeology: a United Kingdom Directory of Resources,** Council for British Archaeology/English Heritage, 1996. ISBN 1-872414-67-2. A comprehensive source book with details of books, audio visual resources, sites, museums and all the relevant addresses.

Hill, J D, Mays, S, & Overy, C, **The Iron Age,** Archaeology and Education No. 9, University of Southampton, 1989, ISBN 0-85432-332-5.

Hodder, I, **The Present Past,** Batsford, 1982, ISBN 07134-2493-1. On the relationship between archaeology and anthropology and how we interpret evidence.

Hodges, H, **Artifacts,** Black, 1964. Out of print but extremely useful on the know-how of ancient technology.

James, S, **Exploring the World of the Celts,** Thames & Hudson, 1993, ISBN 0-500-05067-8. Very thorough, accessible and well illustrated.

Makey, R, **Food for Free,** Collins, 1972, ISBN 1-85052-052-6.

Mears, R, **The Survival Handbook,** The Oxford Illustrated Press, 1990, ISBN 0-946609-88-8.

Reid, M L, **Prehistoric Houses in Britain,** Shire Publications 1993, ISBN 0-7478-0218-1.

Renfrew, J, Black, M, Brears, P, Stead, S, & Corbishley, G, **A Taste of History,** English Heritage/ British Museum, 1993, ISBN 0-7141-1732-3. Recipes and information about food and cooking from prehistory onwards.

* Reynolds, P J, **Farming in the Iron Age,** Cambridge University Press, 1976, ISBN 0-521-21-84-4.

Richards, J D, **Viking Age England,** Batsford/English Heritage, 1991, ISBN 0-7134-06520-4. One of a major series in which leading archaeologists bring the past to life by interpreting historic monuments. Each contains source material for ancient technology projects.

Ross, A, **The Pagan Celts,** Batsford, 1986, ISBN 0-7134-5528-4. Particularly good on literary sources.

* Seymour, J, **The Forgotten Arts,** Dorling Kindersley, 1984, ISBN 0-86318-052-3.

* Seymour, J, **Forgotten Household Crafts,** Dorling Kindersley, 1987, ISBN 0-86318-1740.

Shelter, Shelter Publications, 1973, ISBN 0-394-70991-8. Superb! Available from Compendium Books, London, Tel: 0171 4858944.

Spindler, K, **The Man in the Ice,** Weldenfeld & Nicolson, 1994, ISBN 0-297-81410-9.

Stone, P, **The First Farmers,** Archaeology and Education No. 8. University of Southampton, 1990, ISBN 0-85432-366-X.

Stone, P, & Mackenzie, R, **The Excluded Past: Archaeology in Education,** Routledge, 1994, ISBN 0-415-10545-5. A very important book, which explores in depth some of the educational issues touched on in this volume.

Strandh, S, **A History of the Machine,** A & W Publishers, Inc. 1979, ISBN 0-89479-025-0.

White, K D, Greek and Roman Technology, Thames and Hudson, 1984, ISBN 0-500-40044-X.

Swan, V G, **The Pottery Kilns of Roman Britain,** Oxbow, 1984, ISBN 0-946897-02-6.

Tabor, R, **Traditional Woodland Crafts,** Batsford, 1994, ISBN 0-7134-7500-5.

White K D, **Greek and Roman Technology,** Thames and Hudson, 1984, ISBN 0-500-40044-X.

Wigginton, E, **The Foxfire Book,** Anchor, 1979, ISBN 0-385-7353-4-1295. Available from Compendium Books, London, Tel: 0171 4858944. This is one of an excellent series produced by a teacher and his students who have collected information about the traditional crafts of the Appalachian Mountains, many of which originated in Britain centuries ago.

Wright, A, **Craft Techniques for Traditional Buildings,** Batsford, 1991, ISBN 07134-6418-6.

Shire books on rural history, country crafts and many aspects of archaeology and ethnography. Each compact volume is written by an expert. Just about every project mentioned in this book is covered by this excellent series. Full list available from Shire Publications Ltd, Cromwell House, Church St., Princes Risborough, Bucks HP27 9AJ.

EDUCATIONAL APPROACHES

Durbin, G, Morris, M, & Wilkinson, S, **A Teacher's Guide to Learning from Objects,** English Heritage, 1990, ISBN 1-85074-259-6.

Fairclough, J, **A Teacher's Guide to History Through Role Play,** English Heritage, 1994, ISBN 1-85074-333-9.

Pownell, J, & Hutson, N, **A Teacher's Guide to Science and the Historic Environment,** English Heritage, 1992, ISBN 1-85074-331-2.

Putnam, B & M & Stone, P, **A Teacher's Guide to Using Prehistoric Sites,** English Heritage, 1996, ISBN 1-85074-325-8.

National Curriculum Council, **A Guide for Staff of Museums, Galleries, Historic Houses and Sites,** National Curriculum Council, 1993, ISBN 1-872676-32-4.

Fiction

The most successful fictional accounts of ancient times pay close attention to the technology being used by the characters. Four of the best depict life in the Ice Age, Bronze Age, Stone Age to Roman period and early twentieth century respectively.

* Fidler, K, **The Boy with the Bronze Axe,** 1968, Penguin, ISBN 0-1403- 0563-7.
* Garner, A, **The Stone Book Quartet,** 1992, Harper Collins, ISBN 0-00-184289-7.
* Kurten, B, **Dance of the Tiger,** 1982, Abacus, ISBN 0-349-12121-4.
* Williams, R, **People of the Black Mountains,** 1990, Paladin, 0-586-9058-4.

PLEASE NOTE: A fuller bibliography and list of useful addresses and contacts, including mail order suppliers, is available from the English Heritage Education Service.

Acknowledgements

Thanks to Mike Abbott, Hazel Allen, Carole Amos, Peter Crew, John Farthing, Reg Miles, Damien Sanders, The Earl of Shaftesbury, Nigel Tibbutt, Thomas Wakefield, Jacqui Woods. In excess of 20,000 children with teachers, parents and the staff of the Dorset Outdoor Education Service have helped to develop the projects. Special thanks are due to the children of Cranborne First and Middle Schools and to Chickerell County Primary School.

Opposite :
Ask a 5 year old how people made fire long ago. 'Rubbing sticks' will be the ready answer. Attempting it usually disproves the adage that where there's smoke there must be fire!

Our Education service aims to help teachers at all levels make better use of the resource of the historic environment. Educational groups can make free visits to over 400 historic properties managed by English Heritage. The following booklets are free on request. **Free Educational Visits** contains details of how to book, a list of all English Heritage properties and a booking form. **Using the Historic Environment** is packed with ideas and activities for National Curriculum study and work on site. Our **Resources** catalogue lists all our educational books, videos, posters and slide packs. Please contact:

English Heritage Education Service 429 Oxford Street London W1R 2HD Tel: 0171-973 3442 Fax: 0171-973 3443